DATE			

It Came from Outer Space

IT CAME FROM OUTER SPACE

Everyday Products and Ideas from the Space Program

Marjolijn Bijlefeld and Robert Burke

GREENWOOD PRESS
Westport, Connecticut • London

Library of Congress Cataloging-in-Publication Data

It came from outer space : everyday products and ideas from the space program / Marjolijn Bijlefeld and Robert Burke
p. cm.
Includes bibliographical references and index.
ISBN 0-313-32222-8 (alk. paper)
1. Astronautics—Technology transfer. 2. Academic spin-outs.
TL865.I83 2003
609—dc21 2002192830

British Library Cataloguing in Publication Data is available.

Library of Congress Catalog Card Number: 2002192830
ISBN: 0-313-32222-8

First published in 2003

Greenwood Press, 88 Post Road West, Westport, CT 06881
An imprint of Greenwood Publishing Group, Inc.
www.greenwood.com

Printed in the United States of America

The paper used in this book complies with the Permanent Paper Standard issued by the National Information Standards Organization (Z39.48–1984).

10 9 8 7 6 5 4 3 2 1

Contents

Introduction

Space may be the final frontier, but here on earth, Americans reap the benefits of space exploration every day. Since its founding in 1958, the National Aeronautics and Space Administration (NASA) has worked not only on space exploration, but also on ways to bring the new technologies developed for the space program down to earth. The results have been tremendous. Stronger and more lightweight materials have been developed. Special coatings have added corrosion-resistance, heat-resistance and strength to building materials. Fire-retardant fabrics developed for space suits have found numerous uses. NASA has conducted research into the human body in the unique microgravity conditions of space, which has led to life-saving medical technologies. Infrared imaging systems have been developed to better detect fires, volcanic action, and environmental conditions as well as to conduct surveillance for security.

NASA engineers and researchers have licensed technology to private industries to improve health care, public safety, the environment, computer technology, transportation, industrial and manufacturing technology, and even better home and consumer products.

This volume contains 67 entries on products and technologies that have been either developed or improved upon because of space-related research and technology development. They represent the broad range of products spun off from the space program—from the now-common items such as smoke detectors and fire-retardant materials to the futuristic technologies such as virtual reality and bionic eyes and muscles. There are hundreds more, and scores of new technologies are being introduced all the time. Appendix items contain listings and Web sites of NASA research groups and what they focus on. Each of them has a technology transfer office, to smooth the process of licensing newly developed NASA technologies to the public.

As the space agency moves into the future with its planned missions to Mars as well as sending astronauts to the International Space Station, it faces new technological challenges. As these missions find astronauts farther from home for longer periods of time, it becomes even more challenging to make certain that environmental systems, food supplies, life support systems, and the structure of spacecraft are adequate. No doubt, this new research will generate more improvements and startling new discoveries—not just about our universe, but also about our bodies, our homes, and our Earth.

Overview

Looking through this volume of NASA-related spin-offs—products or technologies that were originally developed by or for the space agency—the sheer variety might be a bit surprising. But the initial surprise wears off when you think about how much overlap there is between space research and what we do on Earth. Transportation, medicine, health, environment, composite materials, safety, strength, miniaturization of parts, flight, better communication systems, and food supply are all heavily researched by NASA and are all a part of our everyday lives.

In NASA parlance they're called success stories, and there are hundreds of examples that spring from each of NASA's research centers: Ames Research Center, Dryden Flight Research Center; Goddard Flight Research Center; Jet Propulsion Laboratory; Johnson Space Center; Kennedy Space Center; Langley Research Center, Glenn Research Center; Marshall Space Flight Center, and Stennis Space Center. Through its technology transfer offices, NASA has been working on moving technologies, either developed for or by the space agency, into the commercial arena.

The only difference now is that the agency is striving even more in that direction. The approximately $13 billion annual budget of the space agency is not just used to study faraway stars and galaxies or explore Mars or push the limits of humans living in space—although it supports all of those projects as well.

NASA's mandate is threefold: to improve life here, to extend life to there, and to find life beyond. It is the first of the mandates that has the most direct bearing on this volume. NASA, through its research divisions and affiliated research programs in academic institutions, works continually to bring space-based technologies to earth.

On April 12, 2002, newly appointed NASA administrator Sean O'Keefe outlined his vision for NASA's goal to understand and protect our home planet. He said:

> We have come to understand that the only way to really comprehend our climate and to protect the scarce resources of our little blue planet is to look at the Earth as a single, whole system. This holistic approach allows us to see how the oceans affect climate on land, for example, and how natural and man-made environmental hazards in one part of the world affect other parts of the world. From the unique vantage point of space we can see, and more importantly, predict, how dust storms in the Sahara will affect crops in the American Midwest. From the unique vantage point of space we can predict how mosquito-borne diseases will spread. From the unique vantage point of

space we can tell a farmer what part of her field needs fertilizer and which part does not.

The mission is to understand and protect our planet. Protection includes using our scarce resources to improve life on Earth by living in an environmentally sound manner. . . .

Protection of our planet also includes changing our transportation systems on Earth so that they are friendly, efficient and environmentally safe. In the future airports will be more efficient, safe travel hubs.

Protection of our home planet includes sharing NASA's unique technology and imagery with other government agencies, academia and industry, to thwart those who seek to do harm or arrest trends that diminish our quality of life.

It Came from Space

NASA has been bringing its technology to earth for years. In fact, some of our more common household products have their roots in the early Apollo missions. Today's cordless tools are cousins to the tools developed so astronauts could collect surface and ground samples on the moon. Clothing specially designed for astronauts has had multiple spin-offs. Durable, lightweight fabrics used for spacesuits are now being used as roof fabric on sports stadiums and shopping centers. A liquid cooling garment designed to keep astronauts cool underneath their bulky moon suits is used by firefighters and people with medical conditions that make them extremely heat sensitive. Technology created to make moon boots that astronauts could walk in comfortably has been used to improve ski boots and athletic shoes. Patient monitoring equipment was developed so physicians on earth could track astronauts' vital signs; similar equipment is standard in hospitals today.

Sunglass and eyeglass lenses often have a scratch-resistant coating—another NASA spin-off developed to protect helmet visors from the harsh space environment. Self-stabilizing rafts, which won't get turned over in rough seas, had their origins in NASA technology.

With the Space Shuttle program, technology spin-offs continued. More than 100 technology spin-offs are directly attributable to this program. The artificial heart and many other miniaturized medical components were created because of miniaturization technology needed for the Space Shuttle program. The thermal protection system—developed to protect astronauts during re-entry into the Earth's atmosphere—now protects race car drivers.

Infrared cameras, diagnostic equipment, and gas detection systems have been developed or improved upon through NASA initiatives. Bar coding on grocery and department store products was developed to allow workers to track the millions of parts used for the Space Shuttle. The rocket fuel that launches Space Shuttles into orbit is also being used to destroy landmines. And special light-emitting diode (LED) lights, developed for plant growth experiments, are being used in a variety of medical procedures.

Medical technology has benefited greatly from the space program. Noninvasive breast cancer screening, less traumatic breast biopsy techniques, magnetic resonance imaging (MRI), cardiac pacemakers and implantable defibrillators, kidney dialysis machines, insulin pumps, and fetal heart monitors are just some of the devices and procedures that were introduced using technologies developed in the space program.

NASA has pushed computer software development into a new realm as well. A search on software development in NASA's technology pages hits on about 3,000 pages—virtual reality software, icing prediction model software, software development projects with private industry, flight software, aeronautics, robotics, communications, and the list continues seemingly without end. In short, nearly every new technological development has required new software development. Software programming and development is the backbone to nearly every NASA project.

Once a NASA technology is spun off into a technology, it often has a ripple effect. Insulating materials developed for NASA have been used in building materials, race cars, and clothing. In other cases, the technology developments bounce back and forth between NASA and industry. For example, NASA researchers developed a device to transmit astronauts' vital signs without them having to be connected by wires to a monitor. The implantable device hasn't yet been used with astronauts, but it is being used to monitor at-risk fetuses. Now that the company is working on miniaturizing the technology further so it can be put into a pill that can be swallowed, NASA is again interested in it to monitor astronauts' health. And sometimes, the first idea for a commercial development gets waylaid as a more obvious or achievable development comes along. A diagnostic probe with electronic sensors developed at Ames Research Center was originally eyed as a way to help neurosurgeons during delicate brain surgery. However, in late 2003 the product is expected on the market as a probe for making an instant decision about whether to send a woman with a suspicious breast lump in for a surgical biopsy.

Many of the products that have their origins in the space program are not stand-alone commercial products today, but are used in a variety of machines. For example, in the early 1980s, a NASA engineer invented the power factor controller. This controller senses the amount of power needed by an electric motor and varies the power according to the need. As a result, power usage is reduced by six to eight percent under normal conditions, and by as much as 65 percent when a motor is idling. The power factor controller has become one of the most widely adopted spin-off technologies. It is used in hundreds of electrically powered systems, including refrigerators, washing machines, typewriters, kidney dialysis machines, and industrial drilling machines.

Space Product Development

To make sure that new technology is brought to market, NASA has a Space Product Development Program and seventeen Commercial Space

Centers working aggressively to bring new and adaptable technologies into the commercial setting. The three main areas of commercial research supported by the Space Product Development Program are biotechnology, agribusiness, and materials research. According to NASA, "The commercial product goals include improved crop development, enhanced refining processes for better fuel efficiency or casting processes that reduce waste, improved drug development, drug production rates, or drug delivery procedures, and use of advanced materials for hip or knee replacements that are more durable and less likely to be rejected by the body."

In addition, these seventeen commercial space centers around the country are often affiliated with state universities and supported by NASA. Each has a particular area of expertise. A key part of their overall missions is to move space-related discoveries and technologies into the private sector. For more information on each of these centers, turn to the Commercial Space Centers section in Appendix F: What's on the Web.

Some of these centers specialize in aerospace engineering, microgravity materials processing, biophysical sciences, turning combustible materials such as fossil fuels into energy, and improving satellite communication networks. The Center for Space Power at Texas A&M University researches new ways to produce power in space. The Space Research Institute at Auburn University has a center that looks at how to make power supplies last longer in the harsh space environment. Others look at materials development, environmental systems, medical informatics, and robotics. The Food Technology Commercial Space Center at Iowa State University researches ways to develop food supplies and systems for missions. Realize that for short missions, it's relatively easy to pack along the food that's needed. But as human space flight missions last months—or even years—it becomes increasingly important to find ways of packaging or growing fresh and palatable food.

Five Enterprise Initiatives

NASA has five major enterprise areas that seek to transfer technology and research developed and conducted by NASA into more common usage.

Aerospace Technology

The goal of the NASA's aerospace program is to revolutionize aviation. The system is nearly at its limits and NASA wants to help develop an environmentally friendly and reliably safe global air transportation system. To improve the environmental impact, the enterprise has as its goal to reduce nitrogen oxide emissions by 70 percent in 10 years and by 80 percent within 25 years and to reduce carbon dioxide emissions of future aircraft by 25 percent and 50 percent during those time frames. Technologies that could help meet those goals include reducing airframe weight and drag, so aircraft use less fuel; develop advanced engines; develop more efficient operations near airports; and develop alternative vehicles and fuels.

To improve safety, NASA has a goal of reducing the rate of aircraft accidents by a factor of 5 within 10 years and by a factor of 10 within 25 years. The agency will also use its data and technology to identify and fix aircraft problems before they occur, work to prevent accidents that recur with some regularity, and reduce the risk of injury if an accident does occur.

Other objectives are to reduce noise, increase the aviation system capacity, and reduce transportation time. In other words, help to develop lighter, quieter, faster, and more efficient aircraft.

While the Aerospace Technology enterprise area seeks to improve aviation here on earth, another primary goal is to advance space transportation by improving mission safety, mission affordability, and expanding the mission reach. The goals here mirror those of Earth-bound aviation: faster, safer and more efficient and affordable spacecraft.

A third goal of the enterprise is to pioneer engineering and technology innovation. For example, NASA reports scientists are working on "a complex new argon-ion laser measurement technique which allows scientists to 'see' sound. With this tool, researchers can very accurately measure turbulence parameters that will help them understand the physics of how a supersonic jet flow creates sounds, enabling more efficient aerodynamic design." Technology innovations may focus on the development of advanced vehicles that use such technologies as "embedded, distributed computing and sensors for vehicle control; active shape control for flight optimization; high-strength carbon nano-tube composite structures; distributed vectored propulsion systems; and self-healing, multi-function materials," according to the Aerospace Technology Enterprise Web site.

Such an aircraft is not too far in the future. On June 3, 2002, NASA issued a press release reporting that a NASA airplane with intelligent flight controls would start calibration missions in 2002 and research missions in 2003. The plane, part of the Intelligent Flight Control (IFC) research project, "is designed to incorporate self-learning neural network concepts into flight control software to enable a pilot to maintain control and safely land an aircraft that has suffered a major systems failure or combat damage."

The enterprise's fourth and final goal is to commercialize the technology developments. The Web site of the Aerospace Technology Enterprise is www.aero-space.nasa.gov.

Office of Biological and Physical Research (OBPR)

This office researches the effects of low gravity on the human body and the environment. According to the enterprise's Web site, it researches two fundamental questions: How can human existence expand beyond the home planet to achieve maximum benefits from space? How do fundamental laws of nature shape the evolution of life?

The office's goals are to enable exploration by conducting research that will allow humans to safely and productively live in space; to test fundamental principles of physics, chemistry and biology in the space environment; to promote commercial research in space, and to use space research to improve academic achievement and the quality of life.

The OBPR brings together physics, biology, chemistry and engineering for its Earth-based and space-based research initiatives. During space flight, humans are exposed to physical and psychological factors not found on earth: increased radiation, reduced gravity, and isolation. The OBPR is working to "design strategies for maintaining health, safety, and performance in the hostile environment of space. In addition to controlling the physical changes that seriously threaten space travelers' health, this Enterprise will conduct research to develop the means for providing crew medical care remotely. OBPR will also conduct research on technology for efficient, self-sustaining, life-support systems to provide safe, hospitable environments for space exploration. NASA will team with other research agencies, the private sector, and academia to establish the scientific foundation for cutting-edge, molecular-scale biomedical technologies for use on Earth and in space," the Web site states.

The International Space Station (ISS) is providing the space agency and this office a particularly effective long-term testing laboratory. Current research includes the effects of microgravity on human and animal reproduction. The ISS will also be accessible to industries interested in space research. NASA has designated 30 percent of ISS resources for commercial use.

The Web site of the OBPR is http://spaceresearch.nasa.gov.

Earth Science Enterprise

NASA's Earth Science Enterprise focuses changes to the global environment, both naturally occurring and brought on by humans. "The vantage point of space provides information about Earth's land, atmosphere, ice, oceans and biota that is obtainable in no other way." Data gathered can be used in the development of environmental policy.

On the Earth Science Web site (www.earth.nasa.gov), earth science is called "science in the national interest... NASA's Earth Science Enterprise develops innovative technologies and applications of remote sensing for solving practical societal problems in food and fiber production, natural hazard mitigation, regional planning, water resources, and national resource management in partnership with other Federal agencies, with industry, and with state and local governments."

In the Earth Science Enterprise's Strategic Plan, the office notes that Earth science has been a NASA goal since the beginning of the space age in 1958. In other words, NASA has been watching the earth for a long time. For example:

NASA launched the first weather satellites in the early 1960s, making possible weather forecasts three to five days in advance.

A decade later, "NASA began to experiment with technologies for remote sensing of land surface features and vegetation, and Landsat became the world's first civilian land imaging satellite. Landsat is now the basic tool for scientific research on regional to global land cover change, helping to resolve such questions as the rate of deforestation in the Amazon and Southeast Asia, and enabling crop yield prediction in the US Midwest by observing the 'greening up' over the growing season."

By the 1980s, the "satellite-based Earth Radiation Budget Experiment and others enabled the study of solar radiation and Earth's absorption and reflection of it to construct the first model of the Earth's energy budget." And during that time, "NASA's Total Ozone Mapping Spectrometer began global monitoring of the annual fluctuations in ozone concentration and distribution, including the growth of the now-famous Antarctic 'ozone hole'..."

Another remote sensing project—the NASA/French TOPEX/Poseidon radar altimeter—allows scientists to watch ocean circulation, monitor progress of El Niño/La Niña formation and dissipation, and allows meteorologists one set of facts on which to generate long-range climate predictions.

All the data collected by the Earth Science Enterprise have a practical effect on our everyday lives. NASA explains it can be used to measure "seasonal and annual variation in soil moisture for agriculture planning and flood hazard assessment"; provide "geospatial information and decision support systems to generate fire hazard maps based on fuel-load and climate conditions for forest and rangeland management"; and provide policy makers a "scientific basis for air quality management decisions."

Human Exploration and Development of Space

This enterprise is made up of the Office of Space Flight and the Office of Biological and Physical Research. The goal of the Human Exploration and Development of Space (HEDS) Enterprise "is to open the space frontier by exploring, using and enabling the development of space. Specifically, the goals of the HEDS Enterprise are to: explore the space frontier; enable humans to live and work permanently in space; enable the commercial development of space; and share the experience and benefits of discovery. The HEDS Enterprise has defined near-term missions as well as those that extend beyond the next 25 years or so.

The HEDS Enterprise timeline described here was created before NASA grounded all Space Shuttle missions after the February 1, 2003, loss of the *Columbia* during its reentry into Earth's atmosphere. All seven crew members aboard died. Consequently, NASA created the Columbia Accident Investigation Board, which was expected to release a final report on the accident in the summer of 2003. While it seems likely that NASA will resume human space flight, perhaps even before the end of 2003, the timing depends on the modifications that must be made to the existing Space Shuttles.

Originally, the HEDS plan called for a continuation of 7- to 14-day Space Shuttle missions and 30- to 90-day International Space Station missions through 2005. Mid-term plans, from 2006 to 2011, call for extending human reach beyond low-Earth orbit and missions that last 100 days. "Lunar missions could also answer questions about how we can use lunar resources commercially and sustain operations at other planetary venues. Establishing space-based and lunar infrastructures will enable new opportunities for the commercial development of space—and hence, more ambitious exploration goals."

Long-term plans, from 2012 to 2025, call for missions that could last from 500 to 1,000 days and extend beyond Earth orbit. During this time, mission designers will learn about the surface of Mars to prepare for future human/robotic missions there and explore the region of space between Earth and Jupiter.

Beyond 2025, the HEDS Enterprise expects missions could last 2,000 days or longer. "As technology advances, human destinations in the outer solar system, such as Ganymede, an ice-covered moon of Jupiter, and Titan, a moon of Saturn that has an atmosphere similar to that of ancient Earth—might become accessible to human missions later this century.

Although unlikely in the coming decades, eventually technology may open the way for major probes to the very edges of our solar system and beyond. These plans are outlined in the Enterprise's Strategic Plan, available on the Web site: www.hq.nasa.gov/osf/.

Space Science

NASA's Space Science Enterprise covers any area of study not covered by the other four Enterprises. As the office states on its Web site, "That just leaves about the whole universe to us." Space science covers everything from the middle levels of Earth's atmosphere to the edge of the universe—including black holes, white holes, wormholes, quasars, galaxies, stars, the sun, and most everything under it. The Chandra X-Ray Observatory and other powerful and new space-watching technologies come under this Enterprise. The Space Science Enterprise Web site is www.spacescience.nasa.gov.

In conjunction with these five NASA Enterprises, NASA's Chief Technologist developed Strategic Technology Areas—"a select group of very advanced technologies that offer the promise of revolutionizing how NASA does business in the future." They are: advanced miniaturization, intelligent systems, compact sensors and instruments, self-sustaining human support, deep-space systems, and intelligent synthesis environment.

A Quick History of Human Space Flight

The Space Age began in 1958 with Project Mercury. It was followed by Project Gemini, Apollo, Skylab, and numerous Space Shuttle missions.

Project Mercury

There were six Mercury flights, totaling only two days and six hours in space. The goal of Project Mercury was in large part to prove that humans could survive space flight and return to Earth alive. Technology developed for Mercury missions included a capsule designed for ballistic reentry, with heat shields to protect the astronaut from the intense heat and radiation. Six astronauts flew in Project Mercury—one crowded capsule. Seven men were chosen as the first astronauts, although Deke Slayton didn't fly due to a heart condition.

Alan Shephard was the first astronaut in space aboard *Mercury Redstone 3* (Freedom 7—each astronaut named his capsule and added the number 7 to represent the original astronauts) on May 5, 1961. The suborbital flight was only fifteen minutes long. Virgil I. "Gus" Grissom was the crew of *Mercury Redstone 4* (Liberty Bell 7) on July 24, 1961. The flight nearly ended in tragedy when water flooded into an opened hatch, sinking the Liberty Bell. Grissom survived to fly again. John H. Glenn, Jr., who would later become a U.S. Senator and the oldest astronaut when he flew aboard the 1998 Space Shuttle *Discovery*, piloted *Mercury Atlas 6* (Friendship 7) on February 20, 1962. During his nearly five-hour flight, he circled Earth three times. *Mercury Atlas 7* (Aurora 7), another five-hour-flight was piloted by M. Scott Carpenter on May 24, 1962; Walter M. Schirra, Jr. stayed in orbit just over nine hours on the *Mercury Atlas 8* (Sigma 7) flight on October 3, 1962. During the final flight of the project, *Mercury Atlas 9* (Faith 7), L. Gordon Cooper, Jr., orbited Earth 22.5 times during 34 hours in space on May 15–16, 1963.

Project Gemini

The goal of the two-man Gemini project was to reach the moon by the end of the 1960s. The astronauts would demonstrate space rendezvous and docking techniques to be used by later Apollo astronauts to separate the lunar lander from the command module and reconnect with it again later. During its ten missions with twenty astronauts (four of whom had flown in Mercury), the Gemini project logged nearly 1,000 hours in space-flight experience. Some of the technological developments of the Gemini program were adding maneuverability to the capsules and the addition of fuel cells instead of batteries for generating electrical power. Gemini also carried the first onboard computers and it was during this project that the first space walks occurred, requiring life support system extensions beyond the interior of the spacecraft. *Gemini 3*, the first flight, was on March 23, 1965. Missions followed each other closely—usually two or three months apart, although *Gemini VI* and *Gemini VII* were both in orbit—and had a rendezvous—in December 1965. *Gemini VII* was on a 14-day mission, the longest space mission until Skylab missions a decade later. *Gemini XII*, the last Gemini flight, lasted from November 11–15, 1966. Astronaut Edwin E. "Buzz" Aldrin, Jr., spent nearly two and a half hours in a successful tethered space walk.

Project Apollo

The goal of the Apollo project was to get to the moon by the end of the 1960s. The first lunar landing occurred on July 20, 1969. Apollo presented a series of new technical challenges; these were the first spacecraft to leave Earth's orbit and orbit the moon, requiring more powerful rockets. The lunar module was designed to fly in a vacuum. After it separated in lunar orbit, it would float to the moon's surface with two astronauts inside. The duration of the Apollo flights also presented challenges: astronauts caught colds and complained about the food. Project Apollo also reminded Americans that space travel was far from routine. A fire inside an Apollo command module on January 27, 1967, killed three astronauts who were training inside. Virgil "Gus" Grissom, Edward White, and Roger Chaffee died. In April 1970, the fate of the three astronauts aboard *Apollo 13* held the world in suspense. An oxygen tank in the service module had exploded and the mission had to abort the planned lunar landing. The crew of James A. Lovell, Jr., Fred W. Haise, Jr., and John L. Swigert, Jr., had to use the lunar module rather than their command module for the return trip. The dramatic events were the subject of a book by Lovell, later made into the movie *Apollo 13*. In total, six expeditions landed on the moon. The first space walk during *Apollo 11* lasted two and a half hours; on *Apollo 17*, moonwalks totaled 22 hours. The final mission in Project Apollo was the Apollo-Soyuz mission, a planned rendezvous between U.S. and Russian spacecraft. It was the first international space mission. In July 1972, the two spacecraft linked for 44 hours and astronauts from *Apollo 18* and cosmonauts visited each other's ships, ate together and exchanged flags and gifts. Project Apollo ended in 1972.

Skylab

Skylab was NASA's space station, designed to let astronauts stay in Earth orbit for extended periods. The actual 100-ton space station was launched on May 14, 1973 aboard the unmanned *Skylab 1*. *Skylab* had two levels, the upper level primarily for conducting experiments; the lower level was more like a living space—with dining room table, three bedrooms a work area, a shower and a bathroom. In the nine months following its launch, three crews of three astronauts spent time aboard *Skylab*, totaling 513 human days in space and conducting thousands of experiments and observations. The empty *Skylab* station reentered Earth's atmosphere and burned up on July 11, 1979.

Space Shuttle

The Space Shuttle program flew its first mission in 1981. The spacecraft itself represented a startling technological advance: unlike the capsules of earlier space missions, Space Shuttles are airplane-like, reusable, spacecraft. Crews on Space Shuttle missions grew larger, from four people on a 1983 flight to eight in 1985. With larger crews and longer missions, there are more opportunities for experiments. Resulting technologies included the Canadian Space Agency's Remote Manipulator System, a large crane built to move large payloads in and out of the cargo bay. A special space walk backpack allows the astronaut to move outside the orbiter with no tether. Some of NASA's darkest hours are connected to the Space Shuttle program—the loss of crew and craft on two Space Shuttle missions. In January 1986 *Challenger* exploded 73 seconds after liftoff. Seven members of the crew, Francis R. Scobee, Michael J. Smith, Ellison S. Onizuka, Judith A. Resnik, Ronald E. McNair, a Hughes employee Gregory Jarvis, and teacher-in-space Christa McAuliffe died.

On February 1, 2003, the Space Shuttle *Columbia* broke up during reentry to the Earth's atmosphere. All seven crew members—NASA astronauts Rick Husband, William McCool, Michael Anderson, Kalpana Chawla, David Brown, Laurel Clark, and Israeli astronaut Ilan Ramon—died. The Space Shuttle program was grounded as NASA and investigators worked to pinpoint precicely what went wrong and how similar accidents could be avoided. A final report by the investigators was expected to be released in the summer of 2003. The three remaining Space Shuttle orbiters are *Discovery, Atlantis,* and *Endeavour.* A June 2003 report stated that the Space Shuttle program could resume as early as December 2003.

Source: Rummerman, Judy A., comp. *Human Space Flight: A Record of Achievement, 1961–1998.* Monographs in Aerospace History #9, NASA History Division. Washington, D.C.: NASA, August 1998.

What's to Come

NASA is now working on its next generation reusable space transportation system to succeed the Space Shuttle. The Space Launch Initiative (SLI) is an agency-wide effort to determine what kind of transportation system will be needed in future missions. While it will take time to design, test, and evaluate a new system, some criteria are known: it will likely have a kerosene-fueled main engine; a two-stage-to-orbit propulsion system, fueled by all kerosene, all hydrogen or some combination of the two; and tested crew escape and survival systems.

NASA is also developing the most sophisticated lab for planetary life support at its Bioregenerative Planetary Life Support Systems Test Complex (BIO-Plex). These systems will be used in the Mars mission scheduled for 2014. A trip to Mars is expected to take six months and an 18-month stay on the planet is anticipated. That means the Mars crew would be gone at least two and a half years, requiring even more planning and preparation for a safe travel and extended stay in space. For both the planned Mars missions and the ongoing International Space Station missions, new technologies will be required and new discoveries are sure to result.

Only in a space-based lab such as the International Space Station, are some experiments possible. For example, NASA has invented a rotating Bioreactor "as a way to study the impact of microgravity on cellular growth both here on Earth and in space," according to a press release. It continues, "Traditional cell-growth research often produces single-cell, pancake-like cultures. The Bioreactor works by spinning a fluid medium filled with cells. The spinning motion neutralizes most of gravity's effects, creating a near-weightless environment that allows cells to grow more freely, in a three-dimensional manner." A private venture, called StelSys, from Baltimore, Maryland, will look at commercializing microgravity research specifically in biological systems. Possible outcomes, according to the press release, include:

> *Biomolecule Production:* Mature liver cells make unique biomolecules for the body. By using the Bioreactor to simulate the natural conditions within the body, we could potentially harvest the biomolecules and use them as a jump start on the road to new drugs or other therapies. This could help us to screen drugs, test them, and get them to patients more quickly.
>
> *Natural Vitamin D3 Production:* People on kidney dialysis need Vitamin D3, but it is expensive to make and difficult to purify. The Bioreactor will allow StelSys to mimic the natural D3 production in kidney cells and assess whether D3 can be produced easily and inexpensively.
>
> *Culturing Infectious Diseases:* Some pathogens that cause disease cannot be grown effectively using traditional cell culturing technology. Use of the Bioreactor could allow us to grow pathogens under conditions similar to those in the body. When scientists have the means to study these pathogens, they may be better able to develop and test treatments for them.
>
> *Liver Assist Device:* Today, people with severe liver failure cannot survive without a transplant. The Bioreactor could lead to the development of a machine to bridge the wait time between diagnosis and transplant, giving hope to the 25,000 Americans who die from liver disease each year.

Research at NASA's facilities here on Earth will continue as well. For example, robotics researchers at the Jet Propulsion Lab and other institutions are working with electroactive polymers (EAPs), which have earned the nickname artificial muscles because they act similarly to biological muscles. The research could one day lead to using EAPs as a bionic replacement for damaged muscles. The impact of such research would be astounding.

In fact, the impact that NASA-related technologies already have could be called astronomical.

Additional Reading

Abboud, Leila. "Looking beyond the Shuttle—Choice of a Successor Craft Will Help Determine Nature of Space Program for Decades." *Wall Street Journal,* 11 February 2003, A17.

"Apollo-era Technology Spinoffs Continue to Enhance Human Life." NASA press release, 13 July 1989.

Gugliotta, Guy and Eric Pianin. "Spaceflight Debate Pits Man vs. Machine: Shuttle Disaster and Unmanned Mission Successes Raise Questions for Congress." *Washington Post,* 27 February 2003, A6.

"Milestone Review Brings NASA One Step Closer to New Launch Vehicle." NASA press release, 30 April 2002.

"NASA Airplane to Health Itself with Intelligent Flight Controls." NASA press release, 3 June 2002.

"New Frontier in Biotechnology" NASA press release, 14 September 2000.

Pianin, Eric. "Flights Likely to Resume, but NASA Itself May Enter New Orbit." *Washington Post,* 2 June 2003, A2.

Reinert, Patty. "NASA Eyes December for Next Shuttle Flight." *Houston Chronicle,* 11 June 2003, A1.

Space Launch Initiatives (to determine the new space transportation system) Web sites. www.slinews.com and www.spacetransportation.com.

"Spinoffs from the Space Program." NASA fact sheet, March 2000.

"Spinoffs from the Space Shuttle Program." NASA fact sheet, March 2000.

Wald, Matthew L. "Future Shuttles May Carry Fewer Astronauts or None at All, NASA Official Says." *New York Times,* 25 March 2003, A10.

Whoriskey, Peter. "Shuttle Failures Raise a Big Question: With a 1-in-57 Disaster Rate, Is Space Exploration Worth the Risk?" *Washington Post,* 10 February 2003, A9.

Airline Deicer—"Ice Zapper"

Ice buildup on aircraft compromises safety. It's not just a cold climate problem, either. Ice can begin forming on wings when airplanes reach their cruising altitude. The combination of low temperatures and speed causes ice to form. Conventional ways to keep ice from forming include thermal deicing techniques and pneumatic boots. Leonard Haslim of NASA's Ames Research Center explained that among the problems with thermal deicers is that they use a lot of energy, they don't keep ice from freezing elsewhere on the aircraft.

In an article in *Aerospace Technology Innovations*, Haslim also explained that "Pneumatic boots inflate slowly and need as much as a quarter inch of ice to accumulate before they start to work. They also dislodge bigger ice pieces that can damage aircraft engines."

Haslim invented the Electro-Expulsive Separation System (EESS), known as the "ice zapper." Haslim, himself a former Naval jet fighter pilot, was concerned about icing on aircraft. Here's how it works.

Two parallel layers of thin copper ribbon are embedded in a rubbery plastic that is bonded to aircraft wings, engine inlets, and other parts where ice formation can be dangerous. The system itself uses a power supply similar to a photoflash. An electric current runs through the copper wiring, creating a repelling magnetic field. That makes the upper copper ribbon jump less than twenty-thousandths of an inch—enough motion to shatter the ice that's building up.

The system can send out these pulses once or twice a minute, throughout the duration of a flight. Ice has virtually no chance to form. Even if ice does form, the ice zapper can shatter ice an inch thick into tiny particles—too small to do damage to the aircraft. The system is also much lighter than conventional methods of deicing.

The ice zapper technology earned Haslim NASA's inventor of the year award in 1998. The technology was licensed to Ice Management Systems, Inc., Temecula, California, in 1995.

The system was first tested on the Lancair IV aircraft in 1998. It took Lancair nearly four years and almost $1 million to bring the EESS patent into commercial production. Thompson Ramo Woolridge (TRW) also purchased the technology for use in a line of unmanned aerial vehicles.

Beyond the aerospace industry, the EESS has potential in the automotive industry as well. A version of the ice zapper could be bonded to car windshields, eliminating the need for ice scrapers. It could also be used on boats.

The ice zapper is certainly not NASA's only foray into deicing. Indeed, ice removal and prevention of ice buildup is so important to NASA that there's an entire Icing Branch at NASA's Glenn Research Center in Ohio. As part of NASA's Aviation Safety Project, the Icing Branch strives to make flight in icing conditions safer through research, education, and partnering with agencies, industry, and academia. It is home to the Icing Research Tunnel (IRT)—a refrigerated wind tunnel. Built at the end of World War II, IRT has been used to test ice protection systems for aircraft. It is still used today to test and develop deicing and anti-icing fluids for use on military and commercial aircraft, as well as for use on the ground. Using water droplets to imitate an icing cloud, the IRT can produce airspeeds up to 400 miles per hour and temperatures as low as −40 degrees Fahrenheit.

See also:

Railroad anti-icer.

Additional Reading

"A Cool Tool for Deicing Planes" *NASA Spinoff 2001* (2001): 73.

Ice Research Branch Web site. http://icebox.grc.nasa.gov.

"NASA Lightweight 'Ice Zapper' to Be Used on New Aircraft." NASA press release, Ames Research Center, 16 June 1998.

"Safer, More Effective, More Efficient Deicer." *Aerospace Technology Innovation* 6, no. 3 (May/June 1998).

Anthrax Protection

Before the fall of 2001, many people had never even heard of anthrax, the airborne pathogen that became synonymous with bioterrorism. Anthrax infection can occur naturally in hooved animals and can be spread to people following contact with infected animals or contaminated animal products, or can occur as a result of intentional release of anthrax spores as a biological weapon. The United States and the Soviet Union are among the nations that have developed biological weapons programs that include anthrax. According to the Center for Civilian Biodefense Studies at the Johns Hopkins University, in theory, fifty kilograms of anthrax released from an aircraft along a two-kilometer line could create a lethal cloud of anthrax spores that would extend beyond twenty kilometers downwind.

Anthrax captured headlines in the United States and worldwide in fall 2001 and worldwide, as five people died from the inhalation form of the disease. Most were mail carriers or worked in office mailrooms and worked in areas through which anthrax-contaminated mail passed. However, the death of two women mystified public health specialists, as they could find no obvious connection between the women and anthrax-tainted mail. Inhalation anthrax is extremely rare; prior to 2001, only eighteen people were known to have died of inhalation anthrax in the United States since 1900. The skin or cutaneous form of the disease is less deadly.

The anthrax-laced letters sent to congressional offices led to shutdowns of those buildings and postal facilities for decontamination. Newspapers, the U.S. Postal Service, and the U.S. Centers for Disease Control all responded by creating

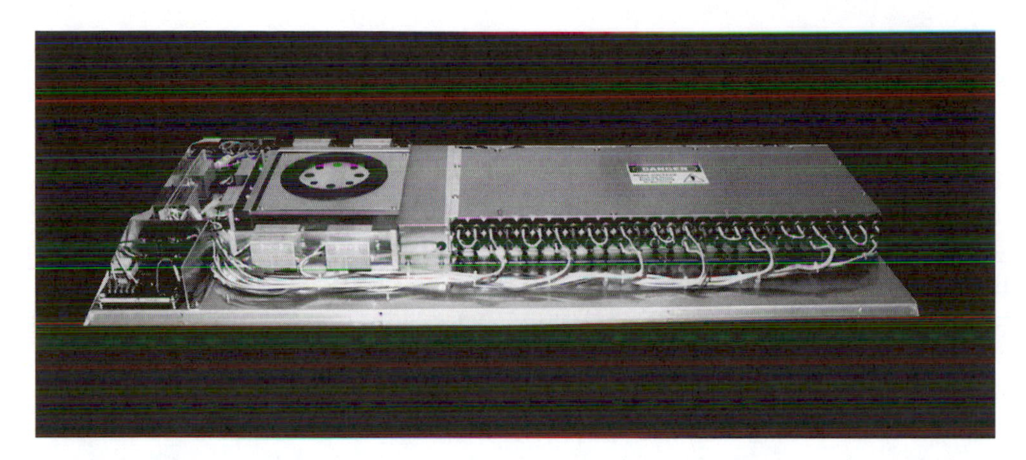

AiroCide TiO_2, an anthrax-killing device made by KES Science and Technology Inc. Photo courtesy NASA.

detailed information on handling suspicious mail. In the midst of the scare, KES Science and Technology of Atlanta, Georgia, was putting the finishing touches on a report concerning a most coincidental finding. It turns out that the company, which had developed with NASA a system to remove ethylene gases from greenhouses to help fruits and vegetables stay fresh longer, could use that same technology to zap anthrax spores. (*See entry* "Fresher Vegetables" for KES's original NASA spin-off technology.)

KES created a system called Bio-KES for use in the produce industry to preserve perishable foods. The system circulates air through a box containing ultraviolet lamps and titanium dioxide (TiO_2)-coated tubes. The combination of the light and TiO_2 convert ethylene into carbon dioxide (CO_2) and water (H_2O). What researchers found was that the Bio-KES system not only removed ethylene gas but also killed airborne dust mites. In May 2000, KES began to modify its Bio-KES system to create an industrial airborne pathogen removal system. Among the changes the company made was adding higher-powered ultraviolet lamps to increase the kill rate. From that, it developed AiroCide TiO_2, a box-like device that is mounted on ceilings. In December 2001, a team of University of Wisconsin professors published a study showing that AiroCide TiO_2 kills 93.3 percent of airborne pathogens, including anthrax spores that pass through it. In addition to the *Bacillus anthraci* or anthrax spores, the AiroCide system also is effective in killing influenza, tuberculosis, *E. coli* and many kinds of airborne viruses. AiroCide TiO_2 can be installed in any common area, such as office mailrooms, conference rooms, kitchens or break rooms.

In the February 1, 2002, article "Annihilating Anthrax," posted on the Web site Science@NASA, the development was detailed. The AiroCide TiO_2 box is about the size of a tabletop. Inside the box are ultraviolet light and bioactive hydroxyl radicals that destroy bacteria, viruses, fungi, molds, and spores. The combination of the light and the ions disrupt the organic molecules. In other words, the spores that pass through the AiroCide TiO_2 device aren't filtered; they're actually killed. In testing at the University of Wisconsin, researchers used a less dangerous cousin of the anthrax spore—*Bacillus thurengiensis*—which acts very similarly. The researchers would direct a cloud of about 1,000 spores through the AiroCide chamber. Approximately 100 came out intact, but since the air in a room keeps circulating, chances are those spores would get zapped on a second run through the box.

The arrangement of TiO_2-coated tubes inside the AiroCide unit is random, so there's no direct path for the air to flow through the machine. The air inside is more turbulent, meaning that the spores stay in the machine longer, increasing the chance that they'll be attacked by the hydroxyl ions and ultraviolet light.

The testing done by University of Wisconsin researchers is available on the KES Web site, www.kesmist.com.

Additional Reading

Lemonick, Michael D. "Anthrax: Lessons Learned." *Time,* 31 December 2001, 126.

McCullough, Marie. "Anthrax Deaths Lead Investigators to Believe That It's Far More Lethal Than Previously Thought." *Knight-Ridder/Tribune News Service,* 27 November 2001.

Phillips, Tony. "Annihilating Anthrax: NASA- and Industry-Sponsored Research Aimed at Growing Plants in Space Has Led to a Device That Attacks and Destroys Airborne Pathogens—Like Anthrax." Science@NASA. 1 February 2002. http://science.msfc.nasa.gov/headlines/y2002/01feb_anthrax.htm. Accessed 6 March 2002.

"RE: Performance of Bio-KES Now Called AiroCide TiO_2 System in Controlling Bacterial Spores." Letter from Dean Tompkins and Terry Kurzynski, Madison, Wisconsin, 1 December 2001, to John Hayman, president, KES Science and Technology. www.kesmist.com. Accessed 6 March 2002.

Anti-Fog Spray

The *Gemini 9* mission in June 1966 was supposed to include a particularly exciting experiment. Astronaut Eugene Cernan was going to spend nearly three hours outside the capsule wearing a jet-powered backpack called the Astronaut Maneuvering Unit and floating alongside the capsule for nearly two full orbits of the Earth.

Unfortunately for Cernan, the experiment never happened. Cernan found working in a near-weightless environment far more difficult than expected. Everything took longer than it had in his training, and the extra exertion caused him to sweat so much that his faceplate fogged up. The problem was particularly acute when the sun shone directly on Cernan; when he and the capsule passed into darkness his fogged visor would slowly clear, but then would become blocked once more when the capsule flew into the light again.

After two hours of slow progress outside the capsule, Cernan and Mission Control agreed to cancel the rest of the experiment. Even then, Cernan's struggles weren't over. He had to grope his way nearly blind along the outside of the capsule back to the cockpit, and by the time he was again inside *Gemini 9* his faceplate was completely fogged again. Fellow astronaut Thomas Stafford said afterward that he couldn't even see Cernan's face, though their helmets were almost touching.

After the mission, NASA engineers duplicated the fogging problem in altitude-chamber tests, using Cernan's space suit. They also began working to develop a new compound to keep faceplates clear. Scientists at Johnson Space Center developed a coating that consisted of deionized water; an oxygen-compatible, fire-resistant oil; and liquid detergent. When they applied the compound to a spot on the inside of Cernan's faceplate and repeated the experiment, the spot stayed clear even when the rest of the mask fogged. The new anti-fog compound immediately became part of

Gemini missions and was applied by astronauts immediately before any space walks.

In 1980, a Florida company, Tracer Chemical Corp. of Tampa, learned about the compound through a NASA publication and used the formula to create a product it sold under its own name and through existing name-brand labels. The company found customers among skiers, motorcyclists, fire departments, and skin divers, as well as at companies making products such as car windows or bathroom mirrors. Today there are many more makers of anti-fog products and the market is even bigger—paintball players, swimmers, and boat and car racing enthusiasts use them.

Cernan's problems during his space walk went beyond a fogged-up faceplate. NASA engineers also had to modify the space suits used by astronauts to give them better protection against the heat of the sun. NASA did eventually test the jet pack that Cernan wasn't able to use, on a Skylab mission in 1973.

Additional Reading

"Anti-Fog Compound." *NASA Spinoff 1985* (1985): 62.

Baker, David. *Inventions from Outer Space.* New York: Random House, 2000.

"Fogless Ski Goggles." *NASA Spinoff 1976* (1976): 90.

Hacker, Barton C. and James M. Grimwood. *On the Shoulders of Titans: A History of Project Gemini.* NASA History Series. Washington, DC: NASA, 1977.

Anti-Shock Trousers

A special kind of trousers worn by test pilots and astronauts from the early days of space exploration up to today has proven to be a valuable tool for paramedics here on Earth. In both cases, the issue is the flow of blood through the body.

During space flight the pull of gravity virtually disappears, dropping to a G-force of 0.001, compared to 1.0 on Earth. Without gravity, blood is no longer pulled down toward the feet, and blood volume in the upper body begins to increase. As astronauts come back to Earth, the full force of gravity takes effect and blood volume in the upper body drops. Crewmembers returning from both short and long missions experienced low blood pressure and "orthostatic intolerance"—the dizzy, lightheaded sensation felt when one stands up too quickly and there isn't enough oxygenated blood flowing to the brain.

The same problem existed back in the early days of aeronautics experiments, so engineers had designed anti-gravity suits for test pilots to help them cope with the G-forces felt during high-performance aircraft flights. The suits used pneumatic bladders that could be inflated to exert pressure on the legs and lower torso, effectively keeping blood volume in the upper body and allowing the brain to get enough oxygen.

In 1969, a physician at Stanford University approached NASA seeking emergency help for a dying patient with internal bleeding that doctors couldn't control. A research team at Ames Research Center, where the space suits for the Apollo mission were designed, provided an anti-gravity pilot's suit, and it stopped the bleeding and saved the patient's life.

Ames researchers continued working with the technology and developed a prototype suit designed to help hemophiliac children from bleeding into elbow and knee joints by straightening and compressing the joint until they receive medical attention. A

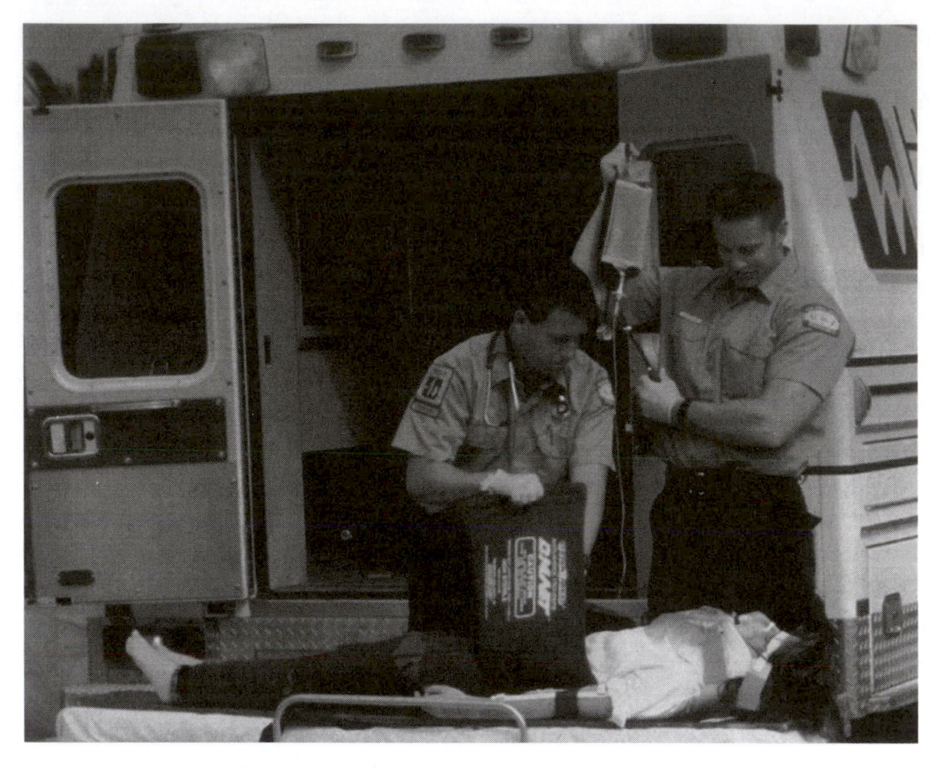

Anti-shock trousers. Photo courtesy NASA.

lack of funding curtailed that research, but a private company, the Zoex Corp., took over development and hired a member of the original Ames team. Zoex developed a suit and obtained a patent for it.

Anti-shock trousers were also widely used during the Vietnam War, and today a number of companies now make similar products. NASA estimates that the technology has been used more than two million times, and that the market for anti-shock garments has topped a cumulative total of more than $50 million. By 1992, more than 75 percent of states required ambulance crews to carry the equipment. In 1997, the American College of Surgeons included anti-shock trousers on its list of essential equipment for ambulances. Anti-shock trouser technology was inducted into the Space Technology Hall of Fame in 1996.

Despite their widespread use, in recent years several studies have suggested anti-shock trousers may not be as consistently effective in helping patients as once thought. They are still considered effective for traumatic hypotension (a severe loss of blood pressure) and in helping to immobilize trauma patients with pelvic fractures or fractures to the femur or lower extremities.

Additional Reading

"Anti-Shock Garment." *NASA Spinoff 1996* (1996): 28.

Dickinson K., and I. Roberts. "Medical Anti-Shock Trousers (Pneumatic Anti-Shock Garments) for Circulatory Support in Patients with Trauma." *Cochrane Review* 1 (2002).

Discovery Health/Space Medicine Web site. http://health.discovery.com/stories/space medicine/pages/saving_lives.html.

"Moving Forward: 1996 Space Technology Hall of Fame." *Aerospace Technology Innovation* 4, no. 2 (May/June 1996).

NASA's Virtual Astronaut Web site. http://virtualastronaut.jsc.nasa.gov/VA_Text_Version/hearttext.htm

Roberts, Ian, Phillip Evans, Frances Bunn, Irene Kwan, and Edward Crowhurst. "Is the Normalization of Blood Pressure in Bleeding Trauma Patients Harmful?" *The Lancet* 357, no. 9253 (3 February 2001): 385.

Athletic Shoe Improvements

Runners and sports players have a certain bounce in their step today because of several NASA-inspired improvements to athletic shoes. Space suit technology has been used in shock absorption, cushioning, the exterior design, and manufacturing of athletic shoes.

Here are some of the developments directly attributable to NASA or the agency's designers and engineers.

Reebok subsidiary AVIA Group International, Inc., of Portland, Oregon, introduced the AVIA Compression Chamber mid-sole in 1990. Some years before, the company had contracted with Alexander "Al" Gross, an aerospace engineer, based in Aspen, Colorado, who had worked on space suit design. The company wanted Gross and his team to develop a better shoe in terms of shock absorption, stability, and flexibility. Gross adapted some ideas from the space suit—which is rigid when it's pressurized but needs to be flexible to allow the astronaut to move around comfortably. According to *NASA Spinoff*, "By layering or combining materials and varying the shape, size and number of bellows, space suit designs can vary joint flexibility." The shoe used the same concept: an external pressurized shell with horizontal bellows for cushioning and vertical columns for stability.

Gross and AVIA also used another NASA technology called "blow molding" to create a more durable shoe. Blow molding was developed originally by NASA to make equipment, such as the lunar helmet and visor, out of a single part without weld lines or cement seams. Seams are the weakest point, so eliminating them makes for a stronger and more impact-resistant product. Blow molding had never been used in the shoe industry prior to this.

Gross has been in demand for his innovative approach to shoe design; the *Denver Rocky Mountain News* notes he's worked for 33 shoes companies. Each year, Gross holds a Sneakerball® fundraising event to benefit at-risk youth. Partygoers must wear sneakers or decorated footwear. An article about the 1999 event in the March 31, 1999, *Denver Rocky Mountain News* said attendees showed up "in glow-in-the-dark sneakers, sequined sneakers and creature sneakers topped with furry stuffed animals. Others were spruced up with faux foliage, bird feathers and holographs."

Another major development in athletic shoes occurred even earlier, as a NASA engineer put the "Air" into Nike shoes. Aerospace engineers Frank Rudy and Bob Bogert had both been involved in space programs. A profile of them in *Machine News* describes how in the 1970s they developed the process of covering pressurized, permanently inflated air beneath the foot to serve as cushioning and shock absorption. Bogert told the magazine, "We had worked on rockets, on guys going to the moon. We thought there would never be another golden era like that in aerospace. We wanted to do something else." Rudy's wife suggested he develop a more comfortable ski boot; Rudy thought of air as the perfect cushioning system. They first developed an inflatable air bladder to fit around the foot and ankle. Because air leaked out, the bladder would need to be pumped up. But air around the foot is different from air under the foot. When Rudy stepped on his liner, pressurized to 25 pounds per square inch (psi) he "realized that the air had much more cushioning value under the foot than around it." The difficulty was in keeping the air cushion inflated permanently.

Although he presented the concept to most shoe manufacturers, only Nike was interested. In 1979, the Nike Tailwind hit the market. To help market the concept, Nike added a "window" so shoe shoppers could "see" the air, actually gases. The shoes were endorsed by basketball legend Michael Jordan and Air Jordan athletic shoes became one of the most popular models Nike has ever sold.

One other NASA-inspired addition to athletic shoes was the Dynacoil system based on a three-dimensional spacer material used for cushioning and ventilation in lunar boots. Athletic shoemaker Kanga-Roos USA (the company no longer exists) used this cushioning system in the shoe's mid-sole. The concept was that where the foot touches the ground, the cushioned fabric would compress, but the fibers would rebound as the pressure lifted, giving a kind of coiled energy spring.

Additional Reading

"Apollo-Era Technology Spinoffs Continue to Enhance Human Life." NASA press release, 13 July 1989.

"Apollo's Contribution to America." *Aerospace Technology Innovation* 7, no. 4 (July/August 1999): 4–5.

Braham, James. "High Tech Afoot." *Machine Design* 63, no. 11 (6 June 1991): 80.

Carr, Debra. "Going the Distance—From Soft-Kid Shoes and Basic Canvas Sneakers to Waffle-Soled Joggers and Air-Inflated, High Tech Designs, the Athletic-Shoe Segment Has Grown from Young Upstart to Slick Marketing and Design Machine." *Footwear News* (26 April 1999): 102.

Sneakerball Web site. www.sneakerball.net.

"Spinoff from a moonboot." *NASA Spinoff 1986* (1986): 84.

Baby Food

Scientists studying ways to feed astronauts during long missions in space couldn't have foreseen what their research would produce. In the 1980s, scientists for aerospace company Martin Marietta were working with NASA to find ways to grow algae as a source for both food and oxygen during manned space voyages. The thinking was that humans spending months or even years in space out of reach of supply ships from Earth would have to grow their own food.

Algae was studied as a potential food for space travel because it is high in nutrients and simple to grow. What algae also offered, though, were two fatty acids essential to brain and eye development in infants: docosahexaenoic acid (DHA) and arachidonic acid (ARA). Both of these fatty acids are abundant in algae and in breast milk. DHA is an omega-3 fatty acid, and is a major structural fatty acid in the brain, and the retina of the eye, as well as in heart tissue. ARA is an omega-6 fatty acid and is the primary omega-6 in the brain. Along with its value in brain development, ARA is a precursor to a group of hormone-like substances called eicosanoids, which help support the body's immune system and blood-clotting ability.

In the late 1970s researchers in England and the United States discovered the links between the fatty acids and brain and eye development, as well as the presence of DHA and ARA in breast milk. They're also found in fatty fish, such as salmon, for which algae is a main food source. Those discoveries, made separately from the NASA-related research, prompted pediatric nutrition specialists to propose putting DHA and ARA into infant formula, for babies whose mothers breastfeed them for only a few months or not at all. However, there wasn't a reliable way to extract the fatty acids from algae. In the late 1980s, though, the scientists who had studied algae for NASA—who were now with a new company, Martek Biosciences of Columbia, Maryland—developed a technique to grow algae and extra DHA and ARA from it. Martek's founder, Richard J. Radmer, was the former director of Martin Marietta's biosciences department. The algae is grown inside seven-story-high fermentation tanks at a plant in Kentucky from natural vegetarian sources. *Crypthecodinium cohnii,* an alga, is a naturally high producer of DHA, while *Mortierella alpina,* a fungus, is a naturally high producer of ARA.

Martek turned its discovery into a product called Formulaid® and began marketing it to the makers of infant formula. The algae-based additive first became available outside the United States. In 1994, the World Health Organization (WHO) recommended the fatty acids be added to infant formula. Also that year, Martek's Formulaid was first used in Belgium in formula made for premature infants, because its presence is especially helpful to infants in the last trimester of development, when a fetus naturally receives DHA and ARA in the womb from its mother. In 1997 an Israeli company added Formulaid to regular infant formula.

In 2001, the U.S. Food and Drug Administration approved the use of Formulaid in infant formulas. Two U.S. companies, Mead Johnson Nutritionals and Ross Products, announced plans to use the additive in their products in the U.S. market in 2002. Today, infant formulas with Formulaid are available in more than sixty countries.

Several studies have shown that adding the algae-based additives can do more than just give infants an essential nutrient—it can actually make them a bit smarter. A 2000 study by the National Institutes of Health, for example, found that babies fed with Formulaid-based formula scored an average of seven points higher on IQ tests than did babies fed with regular formula. The makers of the additive acknowledge the difference is small, but they hope it will help attract parents to the new formula.

NASA actually has a long history in algae-related research. It has studied ways to kill algae in its development of water-purification systems for long-term space travel, and it has studied the development of algae through its Earth Sciences program research into the early evolution of life on Earth, as well as on other planets. Plus, NASA sensors orbiting the Earth have been used to study massive algae outbreaks such as those that appeared along the Atlantic coast following Hurricane Floyd in 1999.

Additional Reading

Chea, Terence. "Seeing Green: For Martek, Turning Algae into Profits Has Been an Evolutionary Process." *Washington Post.* 3 September 2001, E1.

Coghlan, Andy. "Bottle-Fed Babies Need Brain-Boosting Formula." *New Scientist* 150, no. 2030 (18 May 1996): 12.

"Companies Announce Plans to Launch Formula with Nutritional Docosahexaenoic Acid." *Biotech Week,* 30 January 2002, 11.

Dunford, Nurhan T. "Health Benefits and Processing of Lipid-Based Nutritionals." *Food Technology* 55, no. 11 (November 2001): 38.

Martek Biosciences Web site. www.martekbio.com.

Better Braces

Smile when you say NASA. Space agency research and technology has contributed two significant improvements to orthodontia, making braces more effective and less obvious.

First, NASA helped companies Ceradyne, Inc. and 3M Unitek find an appropriate material to create invisible brackets for braces. Transparent polycrystalline alumina, or TPA, is a translucent ceramic that was once used in the nose cones of missiles. It's stronger than steel yet has light-absorbing properties that make it transparent. When used for the brackets onto which orthodontic wires are connected, the braces are much less visible.

The second major development was adapting space wire into braces. A NASA researcher invented a nickel-titanium alloy with unusual properties. One of the first applications of nickel-titanium was as antennae for space capsules. The metal was strong, non-corrosive, and heat-activated; therefore it held its shape better than other wires through temperature changes.

These developments make wearing braces more aesthetically pleasing. Early braces involved attaching bands that ran like train tracks around the teeth. Today, ceramic and clear brackets and these thinner wires make braces almost unnoticeable. Brackets are bonded onto individual teeth and the wires exert a constant gentle pull to align teeth, to close gaps between teeth, or to correct overbites or under bites. Braces are typically worn for one to three years.

As a result of these aesthetic improvements, more adults are wearing braces today. It's estimated that about 1 million adults in the United States and Canada wear braces, about 25 percent more than did in the early 1990s.

For other applications of nickel-titanium, see also:

Memory Metals.

Additional Reading

Brady, Patrick B., "Clinical Applications of Copper Ni-Ti." *Clinical Impressions* (ORMCO Corp.) 4, no. 3 (1995): 8.

Craig, John. "Brace Yourself." *National Post* 2, no. 142 (8 April 2000): 15.

Rayl, A.J.S., with Stephen A. Shoop. "Adult Orthodontics Has Grownups Grinning." *USA Today*, 9 February 2000.

Thornberry, Katherine. "Orthodontic Treatments Go Way Back to the Mummy Years." *Silicon Valley/San Jose Business Journal* 19, no. 28 (9 November 2001): 53.

Williams, Rebecca D. "Perfect Smile Replaces Tin Grin." *FDA Consumer* 29, no. 2 (March 1995): 8.

Biofeedback Improvements

Scientists have been using biofeedback, also called neurofeedback, techniques for decades to help people relax and manage stress. That seemed like a good idea for astronauts, too, for whom small mental mistakes could have disastrous consequences.

But biofeedback was not particularly interesting. Originally, it involved connecting a person to sensors that could graph the individual's responses to particular tones. The idea was to recreate or encourage the relaxation that came about as a result of hearing certain tones.

In the 1980s, psychologist Patrick Doyle of the University of Houston at

Clear Lake was working on biofeedback techniques with astronauts at the Johnson Space Center. Rather than subject the astronauts to the monotony of the existing biofeedback techniques, he developed a series of interactive video games. The first one was Bio-Ball, a baseball game in which it becomes easier to hit the ball when relaxed. With practice, the astronaut was able to relax at will—a skill that could then be transferred to other stressful situations. Bio-Ball was followed by other interactive games, including Bio-Golf and others.

Recently, these games marketed by Creative Multimedia, Inc. have been at the center of an intellectual property dispute between Doyle and the University of Houston.

Nonetheless, the concept changed biofeedback techniques considerably. More recently, the techniques have been applied to games to help children with attention deficit hyperactivity disorder (ADHD). Treatments for the disorder include medications and biofeedback that use the same strategies as described above to reinforce increased attention and concentration. Plus, of course, video games are fun for children.

Researchers at Langley Research Center have used flight simulators to test whether pilots stay mentally engaged during flight. Called the Extended Attention Span Training (EAST), the system provides a more stimulating way to receive dozens of sessions of biofeedback training.

Eastern Virginia Medical School (EVMS) in Norfolk, Virginia, has been testing the system on children with ADHD. Using video games that require focus—such as flight simulation or racing, the signals from sensors attached to a player's forearms, typically, are transmitted to the joystick or other game controller. "As the player's brainwaves come close to an optimal, stress-free pattern, the video game's joystick becomes easier to control. This encourages the player to produce these patterns or signals to succeed at the game," according to an August 10, 2000, NASA press release. The advantages to children with ADHD are numerous: there's no medication, it's noninvasive, it can be monitored remotely, and it's fun.

In describing a test of the system, the NASA press release explained that some of the children were given this new form of biofeedback training while others were undergoing the traditional techniques. The principal investigator of the study said, "The main difference we see between the groups so far is in motivation—the children in the video game group enjoy the sessions more and it is easier for the parents to get them to come to our clinic."

Since then, The Attention Builders, a company based in Richmond, Virginia has created The Attention Trainer a wireless helmet with built-in sensors that interacts with specially adapted software. The system is specifically designed for children aged 7 to 14. The sensors measure levels of attention and transmit this information to special software in the game, allowing the game to provide immediate feedback about a child's attention. While the helmet was not a component in the EVMS tests, the concept was the same.

On the company's Web site, it states, "over time, the child learns how it feels to focus and develops the skill that can help him or her concentrate. For example, in a racecar game, as a child's attention levels go up and down, the software sends signals to the child through the games. The signals are presented through the car's steering, braking, and acceleration. When attention levels increase, the controls become more responsive and the car goes faster."

The company states it takes an average of 40 to 60 sessions to develop better attention skills, and the sessions last between 15 and 30 minutes.

There are other spin-offs from NASA biofeedback research, as well. For example, a *Business Week* story described that biofeedback techniques developed by NASA "to help astronauts handle space sickness are being used to treat people with severe and chronic vomiting."

Additional Reading

The Attention Builders Web site. www.attention.com.

Coates, James. "Darth Vader Lite: A Head Case." *Chicago Tribune,* 13 November 2000.

Guttman, Monika. "Just What the Doctor Ordered." *Family PC* 8, no. 8 (August 2001): 74.

Haskins, Walaika. "All in Your Head." *PC Magazine,* 6 March 2001, 71.

Kirchofer, Tom. "Video Game Reads Player's Mind." *The Boston Herald*, 22 January 2001.

Morris, Bonnie Rothman. "Gadget Tries to Lengthen Young Attention Spans." *New York Times,* 8 February 2001.

"Not Star Trek—But Close" *Technology & Learning* 21, no. 6 (January 2001): 6

"Patient, Heal Thyself." *Business Week,* 16 October 2000, 200.

Pope, A. T. and E. H. Bogart, "Extended Attention Span Training System: Video Game Neurotherapy for Attention Deficit Disorder." *Child Study Journal,* Special Issue on Attention Deficit Disorder, 1996, 26, 39–50.

Rezek, Terry. "Fun and Games through Technology Transfer" *Journal of Aerospace and Defense Industry News,* 23 October 1997.

"Stress Management by Biofeedback." *NASA Spinoff 1997* (1997): 72.

"Video Games May Lead to Better Health through New NASA Technology." NASA press release, 10 August 2000.

Bionic Eyes

The workings of the human eye are remarkable. Light pours in through its clear lens and strikes the retina at the back of the eye, where it is converted to electrical impulses. Those impulses then are transmitted through the optic nerve to the brain to form the images we "see."

The conversion of light to electric impulse is done by millions of rods and cones that fill the retina, a light-sensitive layer that covers about 65 percent of the eye. This conversion is the critical step that makes vision possible. Disease, however, can damage those rods and cones in the retina and cause blindness. Macular degeneration, a condition that damages the retina and most commonly affects the elderly, is one such disease. Another is retinitis pigmentosa, an inherited disease that gradually destroys the retina and the optic nerve.

Medical researchers have been trying for years to discover a way to restore vision by repairing the damaged rods and cones in the retina. They've had little success, largely because they couldn't find a kind of material that could be inserted into the eye without complications. Most of those previous efforts involved silicon-based photodetectors, which don't work because silicon is toxic to the human body and it blocked the flow of nutrients through the eye. Now, a technique developed by NASA-supported researchers has overcome that obstacle. Scientists at the Space Vacuum Epitaxy Center at the University of Houston have discovered a way to create an extraordinarily thin ceramic film that responds to light and could be used to replace damaged rods and cones.

"If we could only replace those damaged rods and cones with artificial ones," Dr. Alex Ignatiev, a professor at the University of Houston and director of the SVEC, said in a NASA interview. "Then a person who is retinally blind might be able to regain some of their sight."

The researchers developed the technology during experiments conducted in space using the Wake Shield Facility (WSF), a 12-foot diameter disk-shaped satellite launched in February 1996 by the Space Shuttle Columbia. The WSF, which was designed by NASA scientists, is valuable because it creates an ultrapure vacuum in low-gravity orbit, which lets scientists grow near perfect and extremely thin layers of film atom-by-atom. That process is called "epitaxy." Researchers are using it to create new kinds of semiconductors that are faster and require less energy than those commonly used in electronic devices.

The Space Vacuum Epitaxy Center is using these ceramic detectors to create artificial retinas that can be implanted in the human eye. The detectors use a thin film of lanthanum-doped lead zirconium titanate (PLZT). The technology is based on the photo-ferroelectric effect—the effect of light on solid materials that have a natural polarization of the electric fields. These artificial devices are extremely small. Each contains 100,000 ceramic detectors that are attached to a polymer film measuring one millimeter by one millimeter. The polymer film is then inserted into the eyeball, and a few weeks later the polymer dissolves, leaving the ceramic retina behind.

When exposed to light the PLZT material emits a small electric current that overlaps the spectral sensitivity of the eye. This electrical current produces a signal that the optic nerve carries to the cortex of the brain, just as a naturally produced signal from a normal eye would. Scientists predict that the brain may find the artificially produced signal unfamiliar but eventually adapt and thus restore sight to the patient.

Additional Reading

Ananthaswamy, Anil. "Eye Strain: To Survive, Bionic Eyes Need to Be Tough as Old Nails." *New Scientist* 173 (9 February 2002): 14.

Cohen, Jon. "The Confusing Mix of Hype and Hope." *Science* 295 (8 February 2002): 1026.

Covault, Craig. "Unusual Shuttle Operations Advance Commercialization." *Aviation Week & Space Technology* 145 (2 December 1996): 27.

Lin, H., J. Wu, and A. Ignatiev, "A Ferroelectric-Superconducting Photo-Detector." *Applied Physics* 80 (1996): 7310.

MD Foundation Web site. www.eyesight.org.

National Eye Institute Web site. www.nei.nih.gov.

Breast Biopsy System

Diagnosing breast cancer through a breast biopsy had been an uncomfortable surgical procedure that resulted in scarring, a week of painful recuperation, and exposure to X-ray radiation. That all changed as a result of NASA's Hubble Space Telescope, launched by NASA in 1990. High-tech silicon chips that convert light into digital images developed for Hubble are now used in breast biopsy systems that allow doctors to biopsy suspicious areas with a hollow core needle instead of a scalpel. And since the imaging technology is more specialized, the woman undergoing the biopsy is exposed to about half of the X-ray radiation of a traditional biopsy.

NASA launched the Hubble Space Telescope to its 300-mile high orbit. However, once it was up in orbit, scientists realized that the Charge Coupled Devices (CCD)—the silicon chips that convert light into electronic images, which can then be manipulated on computer—were not sensitive enough for Hubble's exacting needs. Scientific Imaging Technologies, Inc. of Beaverton, Oregon, developed these new CCDs for researchers at NASA's Goddard Space Flight Center in Greenbelt, Maryland. They doubled the number of photosensors per chip, allowing for better images in low light conditions. In 1997, Space Shuttle *Discovery* astronauts worked on Hubble, installing a Space Telescope Imaging Spectrograph (STIS) with CCDs built by Scientific Imaging Technologies, Inc. That replaced the original Goddard High Resolution Spectrograph and Faint Object Spectrograph originally installed on Hubble. With ten times the resolution of any earth-bound telescope, it provides scientists with precise images and data.

By the time the new CCDs were installed on Hubble, however, the company had already applied the new technology for use in breast imaging equipment for LORAD, a Danbury, Connecticut-based maker of mammography and breast biopsy systems. LORAD, which is an acronym for "low radiation," spotted the potential the technology had for giving physicians more accurate images of breast tissue, thus improving their ability to make a diagnosis.

Approximately 80 percent of suspicious lumps or masses in the breast are benign—not cancerous—but the only way to determine that is through a breast biopsy. When a mammogram reveals a suspicious area, traditionally the woman would undergo surgery in which a tissue sample is removed. By incorporating CCD technology into core biopsy procedures, doctors can pinpoint the problem much more easily. Computers map the exact location of the mass or calcification and can guide the placement of a hollow-core needle to the appropriate spot. The doctor can then extract a small tissue sample through the needle.

It's an outpatient, nonsurgical procedure that leaves a tiny scar because the incision is so much smaller. It's also far less expensive, at about one-fourth the cost of a traditional surgery. When it was first introduced, NASA reports stated the core biopsy cost about $850, compared to a $3,500 surgical procedure. Radiologists at that time estimated the technique could reduce national health care costs by about $1 billion annually.

Compared to a one-week-long recuperation, patients undergoing the core biopsy can walk out of the doctor's office and resume most normal activities. The procedure is done with local anesthesia.

According to an April 27, 1997, Scientific Imaging Technologies, Inc. press release, "This minimally invasive system has been shown to be three to 10 times more sensitive than other breast cancer detection methods, and doctors credit system capabilities with helping save many women's lives." In 1997, Scientific Imaging Technologies was inducted into the United States Space Foundation's Space Technology Hall of Fame for applying the CCD technology to medical imaging.

The National Cancer Institute estimates that one in eight women in the United States will develop breast cancer. A woman's chances of developing breast cancer increase with age. Men can develop breast cancer as well, but the disease is about 100 times more common among women. However, if breast cancer is detected early, the five-year survival rate is 97 percent. Many radiology, medical, and cancer prevention organizations recommend annual mammography screenings and yearly clinical breast examinations beginning at age 40.

Additional Reading

Harwood, William. "Hubble Servicing Complete, Discovery Returns to Earth." *Washington Post*, 22 February 1997, A9.

"Hubble Technology Aids Medical Imaging." *Design News* 52, no. 11 (9 June 1997): 41.

LORAD Corporation Web site. www.loradmedical.com.

"NASA Honoring SITe Backthinned-Technology CCD for Its Role in Combating Breast Cancer: Technology Transfer Helping Save Lives." Scientific Imaging Technologies, Inc. press release, Beaverton, Oregon, 2 April 1997.

"New Biopsy Technique Results from Hubble Technology." *Aerospace Technology Innovation* 2, no. 5 (September/October 1994).

"Oregon-Developed Technology Helping Hubble Space Telescope See: Technology Also Helping Save Lives." Scientific Imaging Technologies, Inc. press release, Beaverton, Oregon, 7 February 1997.

"Remote Sensing Down to Earth." *Aerospace Technology Innovation* 7, no. 5 (September/October 1999).

Scientific Imaging Technologies Inc. Web site. www.site-inc.com.

Burn Treatment

There are more than one million burn injuries per year in the United States and about 3,500 people die from fire- and burn-related deaths due to fires, fiery motor vehicle and aircraft crashes, and contact with electricity or chemicals or hot substances, according to the American Burn Association. There are about 45,000 hospitalizations due to burns; about half of those patients are sent to specialized burn treatment centers.

Ultrasound technology developed by NASA is being used to improve treatment for burn victims. Treatment for severe burn victims used to be a very slow process; physicians would allow dead tissue to come off naturally and then would use skin grafts to close the wounds left behind. Doctors couldn't speed up the process unless they knew precisely how deep the damaged area was. The faster they can determine and remove the dead tissue, either chemically or surgically, the faster the healing can start.

In 1983, scientists from NASA Langley's Nondestructive Measurement Science Branch worked with a team from the Medical College of Virginia (Richmond, Virginia), the University of Aberdeen (Scotland), and the NASA Technology Applications Team (Research Triangle Institute [RTI] North Carolina) to develop an ultrasound system that would provide an immediate skin damage assessment. They adapted NASA's ultrasound technology used to test for flaws or weak spots in aerospace equipment and materials.

The prototype instrument Langley developed—which became the commercial Supra Scanner and was approved by the U.S. Food and Drug Administration in 1990—can measure where burned tissue ends and healthy tissue begins. As NASA explains in its *Spinoff* magazine, "This is possible because, when skin is burned, the protein collagen that makes up some 40 percent of skin becomes more dense. The Langley technique involved directing ultrasonic waves at the burned area. The difference in density between damaged and healthy tissue causes sound waves to reflect at the point of interface."

The system is made up of a scanning transducer and computer, which produce high-resolution color images of up to 37 millimeters, or nearly an inch and a half, beneath the surface of the skin. It "generates cross-sectional images of the skin and provides data on skin surface and subsurface features," according to a fact sheet on the technology. The Supra Scanner contains several exclusive technical features such as activation by voice, a display range of 256 colors, and the NASA depth measurement technology.

Beyond its uses in assessing skin damage due to burns, the scanner can be used in the diagnosis of skin cancer and other skin disorders such as plastic surgery and diagnosis of lymphatic disorders. It is also used to predict wound healing and to monitor the healing process. Supra Scanner maker, Supra Medical Corp. (Chadds Ford, Pennsylvania), is also researching the application of its proprietary ultrasound technology for the noninvasive diagnosis of breast disorders.

Additional Reading

American Burn Association Web site. www.ameriburn.org.

Huber, Lisa. "Virginia Firm Hopes to Use Boost from NASA to Market Cancer Scanner." *Daily Press* (Newport News, Va.), 21 October 1993.

"Saving Burn Victims' Pain and Lives." NASA Fact Sheet, October 1993. http://oea.larc.nasa.gov/PAIS/Burn.html. Accessed 20 March 2002.

"Skin Damage Assessment" *NASA Spinoffs: 30 Year Commemorative Edition* (1992): 39.

"Supra Medical Launches Smaller, Less Expensive Version of Scanner." *Health Industry Today* 57, no. 1 (January 1994): 16.

Child Presence Sensor

Imagine this scene. You're driving home from the grocery store and your baby has fallen asleep in the car seat. When you get home, you start to unload the groceries. Letting the baby snooze a few moments won't hurt, you say. But as you pass through the house, the phone rings. Next thing you know, those two minutes have stretched into ten and the inside of the car is dangerously hot.

According to the non-profit child safety organization Kids 'N Cars, in 2001 at least 34 children died of hyperthermia—overheating—as a result of being left unattended in a car. Others have been injured because they've accidentally put the car in gear causing an accident. There have also been cases of children left alone in cars being abducted or choking.

Three NASA engineers have adapted a technology used on Boeing 757 jets and created an affordable device to alert parents that children remain in the car. The engineers, William "Chris" Edwards, Terry Mack and Edward Modlin, all of NASA's Langley Research Center, have dubbed their device the Child Presence Sensor. As of mid-2002, it was not yet commercially available, but the engineers expected it would sell for about $25 and could be used in any car.

An article on NASAexplores.com, explains how it works. "A sensor in the car seat sends a signal to a receiver attached to the driver's key ring. When the driver takes the keys out of the ignition, if the baby isn't removed from the seat within a short period of time, an alarm sounds. The sensor switch triggers immediately when a child is placed in the seat and deactivates when the child is removed." If the driver moves away from the car, the sensor on the key ring sounds ten warning beeps. If the child isn't removed from the seat within one minute, the alarm beeps continuously.

The sensor is designed to jog the memory of an exhausted or distracted parent who may have forgotten to take the child, or to remind parents that their child is still alone after one minute.

In a NASA press release, principal inventor Chris Edwards, said, "I wanted something that would serve as a second set of eyes and ears, something that could easily and inexpensively be retrofitted to existing child car seats." Edwards has small children of his own and had read about cases around the country where well-meaning parents had inadvertently left a small child in a vehicle with disastrous results.

Edwards knew of a sensor technology developed for the NASA Langley 757 research aircraft. The press release explains, "The aircraft is a highly modified flying research lab for experiments ranging from aviation safety to increasing capacity at major airports. The aircraft sensor is mounted in the main landing-gear area to sense environmental effects acting on the aircraft. That data is then beamed to the cockpit by way of a radio-frequency transmitter and receiver system."

"Co-inventors Terry Mack and Edward Modlin adapted the self-contained radio-frequency technology from the 757 aircraft project and combined it with Modlin's highly sensitive switch technology to create an inexpensive prototype device."

Additional Reading

"Don't Forget the Baby" NASAexplores: 6 June 2002. http://www.nasaexplores.com/lessons/02-041/9-12_index.html. Accessed 7 June 2002.

Kids'N Cars Web site. www.kidsncars.com.

"NASA Develops Child Car-Seat Safety Device." NASA press release, 5 February 2002.

Cold-Weather Clothing

Space is a cold place, where temperatures can drop as low as −148 degrees Fahrenheit. And it's not just astronauts who need protection from the cold. So do space vehicles and the on-board equipment. That's especially true for NASA's Space Shuttle program, for example, because many of the shuttle missions involve work and experiments performed outside the spacecraft.

In its search for an improved insulation material, NASA's Kennedy Space Center in Florida asked a small Massachusetts firm to tackle the problem. The company, Aspen Systems Inc., started with a raw material already considered one of the best insulation materials ever made. The product, called aerogel, is pure silicon dioxide and sand—the same materials as in glass, but aerogel is 99.8 percent air. It is the lightest solid known to man. Guinness World Records awarded that title in 2002 to a piece of aerogel made by Dr. Steven Jones of NASA's Jet Propulsion Laboratory. Jones made an aerogel sample that weighed just 3 milligrams per cubic centimeter. A block of aerogel as large as a human would weigh less than a pound yet would be able to support about 1,000 pounds of weight. It was first invented in 1931 by a California researcher who discovered a way to remove the liquid from a gel-like material without causing it to shrink and lose its characteristics as a solid material. Nicknamed "solid smoke," aerogel in its purest form looks like a piece of translucent foam. It is extremely light, highly porous and has very high thermal insulation value.

Aerogel did not become widely used because it was difficult to mass-produce and extremely fragile. Aspen Systems, though, solved that problem and developed a high-speed and low-cost manufacturing process that was 10 times faster than the previous method. And what it made for NASA was a blanket-like material made up of aerogel-based composites and layers of radiation-shield materials. NASA scientists have experimented with the material for uses such as insulating the Space Shuttle's liquid oxygen tanks and feedlines. Aerogel-based insulation was used on the *Mars Pathfinder* mission, and is part of the 2003 Mars Exploration

Aerogel. Photo courtesy NASA.

Rover mission. Other potential uses, according to NASA, include the Reusable Launch Vehicle, interplanetary propulsion and life-support equipment. Plus, NASA's Johnson Space Center is using the material to develop advanced spacesuit insulation and prototype mittens for Mars exploration.

In 1999, Aspen Systems received Small Business Innovative Research Technology of the Year award from NASA for its work with aerogels. It quickly took the product to market, starting a new company called Aspen Aerogels and creating a new material called Spaceloft™, an advanced version of the material it produced for NASA. Spaceloft is relatively inexpensive, flexible, water-resistant and breathable. In 2001, the Italian apparel maker Corpo Nove began making the first commercial product using Aspen's Spaceloft material. Corpo Nove's 9C jacket can withstand temperatures as low as –50°F.

Besides Spaceloft, Aspen Aerogels is making other aerogel-based products by using the same innovative production method it created for the NASA project. The company makes a product called Cryogel®, a low- to medium-temperature insulation that can be used in refrigerators and to protect pipelines. Cryogel can also be produced as a transparent panel, which makes it perfect for skylights or insulated windows.

Aerogels can protect against heat and fire, too. A product called Pyrogel® made by Aspen Aerogels can withstand temperatures up to 3,000 degrees Celsius. Like the company's other products, it can be made in a variety of forms. As a blanket it can insulate around ovens. When mixed as a composite material it makes a shield for rocket nozzles and Space Shuttle tiles. Flexible Cryogel sheets can be used for clothing worn by firefighters or welders, or to contain heat produced inside electronic devices. The company also makes an aerogel-based insulation called Polar Bear, which is designed for medium-temperature uses such as home insulation, clothing, appliances and automobiles. The market for insulation materials is huge—according to NASA the potential worldwide market for low-cost aerogels is projected to be $10 billion a year by 2005. Plus, the breakthrough that Aspen Systems achieved in producing a better insulation for the space program has the potential to reduce global energy consumption and cut greenhouse gas emissions. As better insulation is used in houses, appliances and other energy-consuming devices, less fuel will be consumed.

Additional Reading

Maskara, Alok. "Almost Nothing Is Taking Off." *World & I* 11, no. 11 (November 1996): 56–64.

Naj, Amal Kumar. "Scientists Close in on Aerogels, Insulation That's Almost as Light as Air." *Wall Street Journal* 225, no. 62 (30 March 1995): B8.

Rummler, Gary. "Blast Off: Students' Basement Tinkering Leads to Space Gel." *Milwaukee Journal Sentinel*, 12 December 2001.

Cool Suits

The surface of the moon is hot—about 265 degrees Fahrenheit during the day. To make sure that Apollo astronauts didn't overheat while exploring the moon's surface or working outside of the spacecraft, NASA engineers needed to find some way to keep astronauts cool. Without some kind of personal cooling system, astronauts outside of the spacecraft were in danger of sweating profusely, and at risk for dangerously rapid heart rates and impaired judgment. So engineers developed a liquid-cooled garment in 1968. The garment is powered by a battery-operated minipump and is worn under the space suit. It has a network of tubes that circulate chilled water throughout.

The need to keep cool in extreme heat isn't just limited to astronauts, of course. The same technology has now been used in a number of settings: medical, military, industrial, automotive and sports.

The first medical application of the suit was in the 1986 case of Stevie Roper, an eight-year-old boy whose body has no natural cooling system because he was born without sweat glands. The rare condition is called hypohydrotic ectodermal dysplasia (HED). The story of Stevie's visit to his aunt in Hampton, Virginia, is detailed in NASA's 30 year commemorative edition of *Spinoff* and on the website of the HED Foundation. While riding in a car without air conditioning, Stevie became dangerously overheated. The family pulled over to douse him with water, and the incident spurred the aunt to action. She contacted NASA's Langley Research Center, which in turn put her in touch with a manufacturer that licensed NASA technology to make cool suits for non-astronauts. The company manufactured a child-size vest and helmet liner, which were able to eliminate 40–60 percent of Stevie's stored body heat and lower his heart rate by 50–80 beats per minute. Now the aunt, Sarah Ann Moody, is founder of HED Foundation, which raises money to buy suits and vests for these children who want to lead a more normal life.

Cool suits have also been used for patients with multiple sclerosis (MS), a chronic, progressively disabling disease of the central nervous system, affecting a patient's thought processes, vision, dexterity, balance and sensation. Decades ago, physicians have found that cooling the body offers temporary improvements in MS symptoms. So cool water baths and exposure to air conditioning have been used regularly. But those cooling methods are not always convenient or comfortable.

But the cool suit lowers the body temperature only by one degree over a half hour. That's not enough to start the physiological reaction of shivering or the constricting of blood vessels, which offsets the effect of the cooling treatment. But it is enough, in many instances, to make an appreciable difference in symptoms. MS patients using the cool suit report better motor skills, speech and thought processes. The effect lasts for about two to four hours after the cooling session.

The suit has also been used for patients who suffer other heat-related illnesses, peripheral neuropathy, epidermolysis bullosa, spina bifida and cerebral palsy.

In 1997, NASA developed a prototype UV garment that is being distributed by the HED Foundation, in agreement with Johnson Space Center's Office of Technology Transfer and Commercialization. These suits are for children with Xeroderma Pigmentosum (XP), a genetic disorder making patients highly prone to developing skin cancer. XP patients must avoid as much as UV exposure as possible. These "NASA" suits include hooded top, pants, face drape, gloves and goggles. With these protective suits, children can venture outside and be part of daily activities that had been off limits to them.

Because these fully UV-protective suits get hot, children often wear cooling vests underneath. Rather than the pump type suit described above, these vests use a "static" cooling system. The vests have special pockets into which cooled gel packs are inserted. The vests stay cool for two to four hours; the gel packs can be removed and are recharged in about 30 minutes in a refrigerator. The suits cost under $2,000. The HED Foundation estimates there are several thousand children around the world with extreme photophobia, sun sensitivity or other conditions that demand they stay inside most of the time.

The NASA personal cooling technology is used by several manufacturers for many different types of applications. One contractor developed a liquid-cooled helmet liner for military pilots "after a series of accidents in Vietnam had suggested heat exhaustion as the cause," according to NASA. Pilots or crewmembers of unpressurized aircraft and service people stationed in hot regions are obvious candidates for personal cooling technology.

Other applications include vests or suits for workers in mines, nuclear power plants, steel mills or workers handling hazardous materials who need to wear heavy protective suits for long periods of time. Firefighters have also used the technology.

One other spin-off use has been for race car drivers. A February 2002 article in *Circle Track* speculated that personal cooling systems for race car drivers could become mandatory equipment. The article cites NASA's finding that "when the astronaut's core temperatures went up as little as 1.5 degrees Fahrenheit they made up to eighty unrecognized mistakes an hour." For race car drivers, even little mistakes could result in serious injury to themselves or others around them.

Cooling technology also has some applications for athletes. Wraps using this type of cooling technology are less bulky than ice packs.

Additional Reading

"Cool Suit." *NASA Spinoffs: 30 Year Commemorative Edition.* (1992): 46.

"Cool Suit." *NASA Spinoff 1987* (1987): 104.

Frazier, Meghan. "Don't Be a Hot Head: A Look at a Variety of Cooling Systems Ranging from a Basic T-Shirt to a Miniature Air-Condition Unit for the Car." *Circle Track* 21, no. 2 (February 2002): 59.

Life Enhancement Technologies Web site. www.2bcool.com.

"NASA and Multiple Sclerosis Association to Collaborate." NASA press release, 23 May 1994.

Sayre, R. M, and J. C. Dowdy, J. Stanfield, J. M. Menter, K. L. Hatch, and B. L. Slaten. "Evaluation of the 'NASA' Garment for Children with Xeroderma Pigmentosum." http://www.hedfoundation.org. Accessed 7 June 2003.

"Spacesuit-based Garment Gives Children Days in the Sun." Johnson Space Center press release, February 1999.

"Spinoff from a Moonsuit (Cool Suit)." *NASA Spinoff 1989* (1989): 56

Suriano, Robyn. "Spacesuit Gives Teenager the Ultimate Sunscreen." *Orlando Sentinel*, 3 July 2001.

"Two Special Girls Use NASA Spacesuit Technology to Finally Have Their Day in the Sun." Johnson Space Center press release, 23 April 1999.

"UAH Nursing Conference Shows How 'Cool Suits' Help Kids Sheltered from the Outdoors Lead More Normal Lives." University of Alabama at Huntsville press release, 18 March 1999.

Cordless Tools

When NASA's Apollo astronauts first landed on the moon, there was more for them to do than simply leave footprints and an American flag. Moon landings presented scientists with a phenomenal opportunity to learn more about the moon's geology by studying samples of lunar rock and soil collected by astronauts.

The problem was how to gather those samples. Astronauts were able to scoop up rock and soil from the surface, but they also needed a drill that could bore as much as ten feet below the lunar surface to gather core samples from the lunar crust. Besides being lightweight, the drill had to have its own power supply so astronauts could use it far from the lunar module to collect a wider range of samples.

For the early Apollo moon landings, which began in July 1969 with the historic *Apollo 11* mission, astronauts were restricted to scooping moon soil and collected rocks with shovels. To gather samples from beneath the surface, they had to hammer tubes into the lunar soil. The bulky, pressurized suits astronauts wore made it difficult to do much digging by hand. Alan Bean, who flew on the *Apollo 12* mission in 1969, noted the problem, according to a NASA transcript of the crew's post-mission debriefing: "We had a shovel that we used for trenching; but, because of the length of the extension handle and the inability to lean over and what have you, we could never trench more than about eight inches. That was about the best we could do, and that was a pretty big effort. If we're going to do any good geology, it's going to take a lot of trenching to get down below the surface."

A portable drill was even more essential for the *Apollo 15* mission, which represented a major increase in NASA's lunar exploration mission. The three moon missions thus far had been mostly to test the space program's ability to land a manned spacecraft on the moon and return it to Earth. On the *Apollo 15* mission, though, astronauts were making first use of the battery-powered Lunar Roving Vehicle. It was able to carry two astronauts and their equipment as much as five kilometers from the lunar module.

As part of preparation for the lunar exploration, NASA asked the Black & Decker Manufacturing Company, based in Towson, Maryland, to develop a tool for the job. Black & Decker had developed the world's first cordless electric drill in 1961. For the NASA project, Black & Decker engineers used a newly developed computer program to create a more powerful and more efficient magnet motor that could squeeze every ounce of energy from the attached battery pack.

The first drill was used on the *Apollo 15* mission in the summer of 1971. The lunar drill used by mission commander David R. Scott was nothing like the cordless tools of today. It looked more like a jackhammer and was about the same size, with two handles at the top on either side of a box-shaped battery that drove the electric motor. Astronauts attached 53-centimeter drill-stem sections one at a time as they bored down into the lunar surface to gather samples and insert data-gathering probes.

Working the drill was not easy during its first mission. Visitors to the NASA History Office Web site can watch a video of Scott struggling to push the drill down in the moon's one-sixth gravity. A problem with the design of the drill probes—they didn't carry cuttings out of the hole and jammed—was fixed for later Apollo missions.

For Black & Decker, though, the experience was fruitful. The company took what it learned while developing the lunar drill and began developing a line of battery-powered devices with the same portability that made the lunar drill so handy. The result was a very successful line of consumer products, such as the DustBuster portable vacuum, first introduced in 1979. Later came electric screwdrivers and drills and an assortment of gardening tools, which all owe their start to the needs of NASA's moon exploration program. The battery that Black & Decker developed to power the devices is its VersaPak rechargeable battery system. It uses 3.6-volt nickel-cadmium batteries that are about the size of a hot dog and are able to handle more than 300 rechargings. Nickel-cadmium batteries have been available since the early 1960s.

Cordless tools continue to be part of NASA's space missions. Cordless tools, including a cordless wrench designed by engineers at the Goddard Space Flight Center in Greenbelt, MD, have been used to assemble the Hubble Space Telescope and the $27 billion International Space Station. The cordless wrench is an uncommonly sophisticated tool—it uses a microprocessor to control torque pressure and the turn angle. It can even be programmed so that no fasteners are overtightened; operators get feedback through LEDs and an alphanumeric display.

According to market analysis by the Cleveland, Ohio-based Fredonia Group, consumers clearly prefer cordless tools. Consumer spending for cordless tools is expected to grow 8.4 percent a year through 2004 when it will reach $700 million. By comparison, demand for corded tools is expected to rise less than 2 percent over the same time period.

Additional Reading

Brack, Ken. "Power Surge: Strong Demand Continues for Higher-volt Cordless Tools and Top-performing Batteries." *Industrial Distribution* 88, no. 8 (August 1999): 75.

Caminiti, Susan. "A Star Is Born." *Fortune,* Autumn/Winter 1993, 44–48.

Gunther, Judith Anne. "Unplugged!" *Popular Science,* September 1996, 68–72.
Halverson, Richard. "Retailers Cut the Cord on Low-voltage Power Tools." *Discount Store News* 35, no. 10 (May 20, 1996): 35–38.
"Marketbuster: The Vac That Roared." *Time,* 11 February 1985, 72.
"Spinoff from a Moon Tool." *NASA Spinoff 1981* (1981): 74.
Weber, Austin. "Just Add Batteries: Cordless Tools Offer Many Advantages for Assemblers." *Assembly* 44 (November 2001): 28.

Digital Nose

Maybe you've heard someone use the phrase, "I smell trouble." That saying underscores one of the most important functions of the sense of smell—its ability to detect danger, often before we can see it.

Animals use it to avoid predators, and now astronauts are using olfaction—the sense of smell—to sniff out problems in the fragile environment of a spacecraft. Researchers at the California Institute of Technology, supported by NASA's Jet Propulsion Laboratory, developed a technology thousands of times more sensitive than the human nose, and use it to monitor air quality during NASA missions in space. That's an important role because the slightest leak or spill or spark inside the spacecraft must be dealt with quickly.

Here's how it works: Caltech researchers created a kind of polymer sensor that swells up when exposed to odor molecules. Then they coated the sensors with carbon black, a material produced by the incomplete burning of oil or gas. The carbon acts as an electrical conductor. When the sensors swell up, it pushes the carbon molecules apart, which changes the resistance of the sensor to electricity and produces a different pattern of current. Since different odors produce different kinds of swelling in the sensors, each one also produces a unique pattern of current—a smellprint—that the device can be programmed to recognize. The system works quickly, taking anywhere from milliseconds to a few seconds to recognize a compound.

Digital nose. Photo courtesy NASA.

In 2000, NASA began using the technology in an air-quality monitoring system used during Space Shuttle missions. That same year, a California company called Cyrano Sciences licensed the Caltech technology and went one step further, developing a hand-held "digital nose" called the Cyranose 320. (Both the company and its product get their name from Cyrano de Bergerac, the 17th century soldier and hero of a play by Edmond Rostand who was famous for his large nose.)

The Cyranose is a handy tool in a number of fields, the company says. Food processors can use it to quickly test the freshness of raw materials, such as cheese or spices, and avoid the losses caused when spoiled goods make their way into the final product. Manufacturers can use it in the production of plastics, gasoline and detergents—anywhere they need to monitor a product's chemical integrity. It can also be of use in hazardous situations: it allows emergency crews to quickly evaluate the nature of a hazardous-materials accident and to choose the right equipment and containment strategies.

The Cyranose 320 weighs less than two pounds and looks like a walkie-talkie. The part that looks like an antenna is actually a six-inch sensor used to detect odors (the company calls it a "snout"). Users can customize the device for a particular application the first time they use it. In the initial training session the user calibrates the device by exposing the sensor to samples that it will encounter during regular use. Then the device will respond when it detects an odor that produces the same electronic pattern, or smellprint.

Running a test takes about a minute; the user simply holds the device near the material to be tested and presses a button. The company has also begun exploring ways to use the technology in medicine. Researchers at the Children's Hospital in Los Angeles are using the technology to diagnose upper respiratory infections. At the University of California, Los Angeles Dental School, it's being used to study the bacteria that cause oral malodor—also known as bad breath. In future applications, it could be used as a noninvasive way for doctors and dentists to diagnose infections, liver and kidney disorders and metabolic problems. And after the well-publicized anthrax attacks of late 2001, the company began tests to see if it can detect the odor of the bacterium that causes anthrax.

These medical applications are part of the development of even smaller products using the NASA-developed technology. Cyrano Sciences says it plans to use the same technology to make household devices, such as room air-monitors and appliances that can tell you when food is spoiled. The Cyranose 320 sells for about $8,000.

Additional Reading

"Cyrano 'Nose' the Smell of Success." *NASA Spinoff 2001* (2001): 72.

Cyrano Sciences Web site. www.cyranosciences.com.

"The Cyranose Chemical Vapor Analyzer." *Sensors* 17, no. 8 (August 2000): 56.

Lemley, Brad. "Future Food." *Discover,* December 2000.

Philadelphia, Desa. "A Nose for Anthrax?" *Time,* 26 November 2001.

Rosch, Winn. "Nosing Out Terrorism." *ExtremeTech,* 10 February 2002.

Salkever, Alex. "Sniffing Out Bioterror Threats." *Business Week,* 22 October 2001.

Doppler Radar—Helping Aircraft Avoid Danger

Microbursts are invisible but deadly traps for aircraft that for years were barely understood. They are columns of air, cooled by thunderstorms, which plummet downward at speeds up to 150 miles per hour. Here's why they're so dangerous: pilots flying into a microburst first experience a strong wind blowing toward them, which increases the lift of the plane and causes the pilot to reduce engine power in order to maintain a given altitude. (Planes flying low and slow, such as during takeoffs or landings, are particularly vulnerable.) But when the aircraft passes through the downdraft to the other side, the wind direction suddenly blows toward the aircraft from behind, causing it to lose aerodynamic lift because wind speed over the wing is slowed. Before the pilot can increase engine power, the plane begins to rapidly lose altitude and crashes. Oftentimes planes that crash due to microburst hit the ground tail-first, with the nose up, which indicates that they were trying to fly but weren't going fast enough to create lift.

Despite the threat, little was known about the phenomenon until the mid-1970s, when University of Chicago professor Tetsuya T. Fujita, creator of the F-scale for measuring tornado severity, first coined the term after investigating an airplane crash and seeing starburst-shaped patterns in the damage to objects on the ground. In a severe thunderstorm, for example, all the trees would be knocked down in the same direction. But in a microburst, objects on the ground are blown over in an outward pattern from a central point.

In August 1985, a Delta Airlines Lockheed L-1011 jetliner with 137 people went down when it encountered a microburst during a thunderstorm near the Dallas-Fort Worth Airport. That tragedy triggered a new push by the Federal Aviation Administration to prevent future crashes, and led to the development by NASA of new technology to let pilots see microbursts.

NASA's aerospace engineers at the Langley Research Center in Virginia developed forward-looking radar based on the Doppler effect. That refers to the effect of movement on the frequency of sound and light waves. This is often demonstrated by the sound of a passing train whistle—it sounds higher in pitch as the train approaches and lower as it moves away, because there are more sound waves traveling toward the listener as it approaches and fewer as it moves away. If the whistle weren't moving, a listener wouldn't hear any change in pitch because the number of sound waves would stay constant. It is named for Christian Andreas Doppler, an Austrian physicist who first described the phenomenon in 1842.

The NASA-developed system bounces radio waves off of the tiny droplets that are swept along by wind during a storm. When the waves return to the plane, instruments measure the difference in their speed, which tells pilots the wind speed in the air mass ahead of them. Langley engineers essentially started from scratch in developing the new technology. They spent five years developing prototype systems and then outfitted a Boeing 737-100 jetliner with a windowless "second cockpit" filled with testing gear. Over two years, the Langley team few 130 flights and encountered 75 microbursts. The flights tested the ability of the NASA sensors to measure wind velocities two to three miles ahead of the aircraft and gave pilots 20 to 40 seconds of warning.

Langley's ability to produce the new technology was borne from its expertise in aviation and aerospace vehicle systems technology, as well as atmospheric research. The first commercially produced system for wind shear detection was the RDR-4B forward-looking radar, made by AlliedSignal Commercial Avionics Systems of Fort Lauderdale, Florida. The device was based on technology developed at Langley, and made its first flight in November 1994 aboard a Boeing 737-300 jet flight from Washington, D.C. to Cleveland, Ohio.

Warning systems are now required on all new aircraft, and many older planes have had the microburst-detecting equipment installed as well. Doppler-based systems have also been installed on the ground at airports around the country.

Additional Reading

Curran, Lawrence and Tobias Naegele. "How the FAA and NASA Plan to Combat Wind Shear." *Electronics* 62, no. 33 (January 1989): 33.

Nordwell, Bruce D. "Modified Doppler Detects Wind Shear More Reliably." *Aviation Week & Space Technology* 137, no. 10 (7 September 1992): 143.

"Technology for Safer Skies." *NASA Spinoff 1995* (1995).

Ergonomic Chairs

An unusual thing happened in SkyLab in the 1970s. Crewmembers noticed that when they relaxed, their bodies assumed the same position—about a 128 degree angle from trunk to thigh. It was clearly different from posture on Earth. When we sit in chairs, for example, our bodies are more upright because of the effects of gravity. But the zero-gravity posture is a more natural one. The vertebrae are correctly aligned and the stress on the skeletal and muscular systems is less.

Scientists at Johnson Space Center began measuring these anthropomorphic changes noticed in SkyLab and Space Shuttle Missions and published them. The resulting source book became the inspiration for the BodyBilt ergonomic chair.

The Navasota, Texas-based company designed ergonomic chairs that emulate this zero-gravity posture. Each chair has a contoured seat and numerous adjustable controls to customize the fit. BodyBilt has created ergonomically correct chairs for a variety of uses. There are office chairs, boardroom or conference room chairs, executive chairs, medical seating, and guest seating.

With one touch, the user can put the chair into the natural, stress-free "zero-gravity posture" position. BodyBilt chairs are being used by national clients such as IBM, Caterpillar, Boeing, Intuit, Lockheed-Martin, Hewlett Packard, Level 3, Ernst & Young, Pulitzer Publishing, Lucent Technologies, Anheuser Busch, Avon, Sandia National Labs, United Airlines, and the *Washington Post*, among others.

The BodyBilt chair received an enormous publicity boost when attorneys on both sides of the O.J. Simpson trial in 1995 bought or borrowed a total of 10 of the chairs to alleviate back pain. O.J. Simpson, however, as defendant, did not get one. The company said the interest led to sales of about 6,000 chairs.

Comfortable seating is not just a nicety; it can be a necessity. Sprains and strains, most often involving the back, accounted for more than four out of ten injuries and illnesses resulting in time away from work, according to the U.S. Department of Labor (DOL). Repetitive motion injuries, as a result of grasping tools, scanning groceries and typing, "resulted in the longest absences from work among the leading events and exposures—a median of 19 days. The median days for this event had steadily declined from a high of 20 days in 1992 to a low of 15 days in 1998 before increasing to 17 days in 1999," according to a DOL press release.

On its Web site, BodyBilt also emphasizes that ergonomic chairs not only reduce the risk of injuries, pain, and strain but also improve overall productivity. In normal seating, it says, "As the day wears on, [the employees'] productivity declines, until they leave, worn out, at the end of the day. Ergonomic equipment

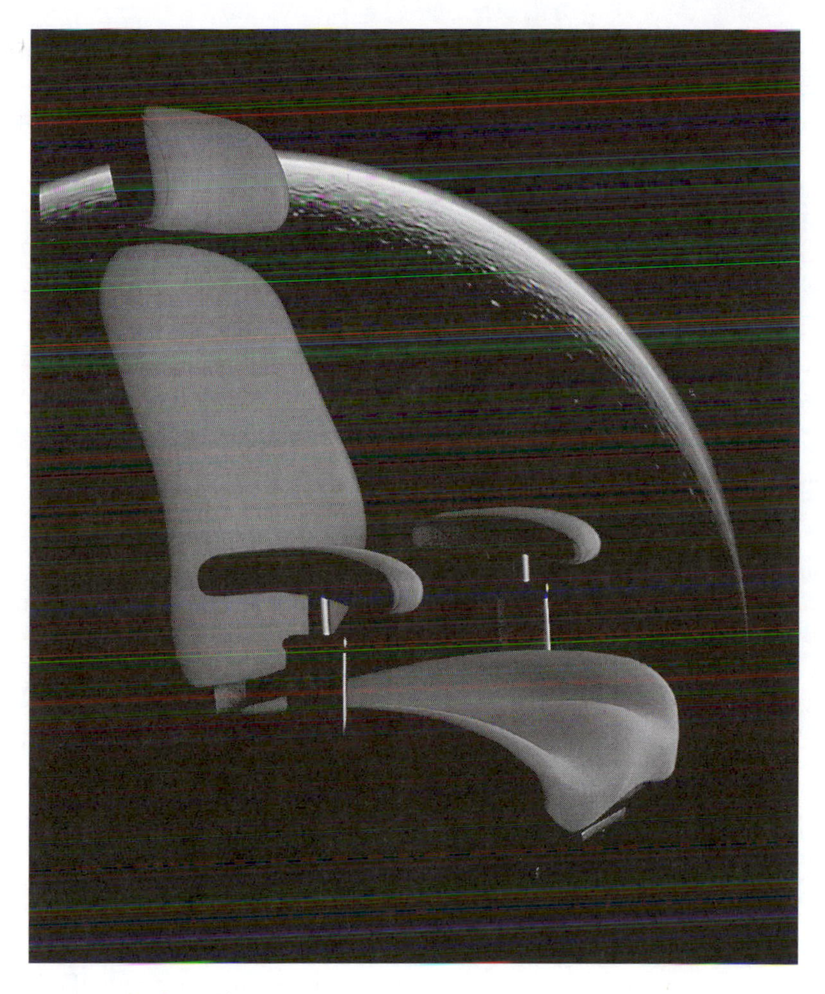

Ergonomic chair. Photo courtesy NASA.

enhances productivity by slowing that decline in productivity, which gives you an average increase in output throughout the day. Increased output usually translates to increased dollars, and the payback on the ergonomic equipment can be figured accordingly."

BodyBilt chairs were used in a test on worker productivity at an Internal Revenue Service office in Austin, Texas. In two studies, permanent, experienced data entry operators had their regular chairs replaced with BodyBilt chairs for a 50-hour week. Comparisons between the weeks showed an average increase in productivity of 8 percent during the time the operator was using the BodyBilt chair. Other factors might have been at play, but the workers themselves said they were more comfortable and were able to work longer without needing to take a break to stretch and move around. In addition, the study organizers expected that since 40 percent of sick leave is attributable to neck, shoulder and back discomfort, the office could expect a reduction in sick leave as a result of these ergonomic chairs that mimic the zero-gravity posture.

Additional Reading

Adhikari, Richard. "Do Vendors Feel Your Pain?" *Information Week,* 4 March 1996, 44

Bernheim, Daniel. "Why O.J. Didn't Get the Chair." *Los Angeles Magazine,* October 1995, 22.

BodyBilt Web site. www.bodybilt.com.

"Ergonomic Chairs." *NASA Spinoff 1997* (1997): 74.

Goldsborough, Reid. "Finding a Good Computer Chair." *Consumers' Research Magazine,* June 2000, 32.

"Lost-Worktime Injuries and Illnesses: Characteristics and Resulting Time Away from Work, 2000." U.S. Department of Labor press release, 10 April 2002.

"Make Yourself Comfortable: Rule No. 1: Don't Just Sit There!" *Consumer Reports,* September 1996, 34.

"Sittin' Pretty: Chairwise, the O.J. Lawyers Have Already Rested Their Cases." *People Weekly,* 25 September 1995, 95.

"Sitting Pretty—And Pain Free" *Business Week,* 25 October 1999, 194

Westbrook, Grant. "Science Can Prevent Injury." *The Business Journal* (Raleigh, NC), 29 June 2001, 22.

Young, Jeffrey. "For That Floating Feeling." *Computer Life* 2, no. 9 (September 1995): 45.

Fire Safety

From a smoke detector to high-tech fire fighting gear, some of NASA's most significant contributions to public safety have been the areas of fire prevention and fire fighting.

On January 27, 1967, astronauts Virgil "Gus" Grissom, Edward White, and Roger Chaffee died in a fire during a training session inside an Apollo command module. As a result of the tragedy, NASA redesigned the spacecraft and redoubled its research into fire fighting and fire prevention strategies in space. Since then, NASA technologies have led to fire-retardant materials, home smoke detectors, better masks, better breathing systems for firefighters, and devices to help fire fighters "see" through smoke and also identify hot spots that could flare up into new fires.

In astronauts' clothing, for example, Monsanto Company developed a chemically treated fabric called Durette, which will not burn or produce noxious fumes. It was used for Apollo astronaut suits and has since been used in a variety of applications, including race car drivers' suits, and in bags for filtering gases and dust from boilers and electric generators.

In early space missions, detecting a fire on board would not have been too difficult. The cabins were tiny and the entire interior could be seen by the astronaut. However, as spacecraft became larger, the need for an accurate smoke and fire detection system became more critical. For *Skylab,* America's first space station which orbited in the mid 1970s, Honeywell Inc. of Minneapolis, Minnesota developed a sophisticated smoke and fire detector. Honeywell Consumer Products incorporated the technology to make home smoke detectors. Since then, smoke detectors have been credited with saving thousands of lives and are required by law in new homes.

Smoke detectors work on the same principles on Earth as they do in space. However, their placement is different. Here on Earth, smoke and heat rise, so smoke detectors are generally installed on ceilings or above doorways. But in space, smoke doesn't rise. Smoke detectors there are installed in or near the ventilation system.

In 1997, another product came onto the market to help fire fighters "see" flames through smoke. Called the FireScape by SafetySCAN, the portable binocular-like device helped the fire fighter detect invisible alcohol and hydrogen flames and visible smoldering embers obscured by smoke, rain, and fog. Prior to FireScape, fire fighters would hold a straw broom over suspect areas. If the broom burst into flames, an invisible fire was burning.

Also in 1997, two Houston firefighters visited the NASA Johnson Space Center (JSC) Technology and Transfer Commercialization office, bringing with them a badly damaged fire helmet. The design of the leather helmet had not changed much since it was developed back in the 1800s. So JSC started working with the Houston Fire Department, the Department of Defense, and Lockheed Martin to develop a prototype suit that would allow a firefighter to be better protected and stay cooler longer. The concept of the suit includes several NASA technologies: liquid cooling (*see* "Cool Suits"), greater impact resistance, protection against hazardous materials, and greater skin coverage. Helmets for future fire suits could include temperature sensors and infrared imaging, which would help the firefighter search for victims.

The National Fire Protection Association and NASA's Marshall Space Flight Center have also been working on technology transfer ideas to improve fire safety. Some of the ideas being worked on include monitors to gauge structural integrity, personnel locator systems so that the emergency crew in a building can be tracked, vital signs monitors so that workers could be warned to get out of potentially dangerous situations, hazardous materials sensors, and thermal sensors that could detect if there is a fire behind a wall.

These types of collaborations are not unusual. After four firefighters died on the job in Pittsburgh in 1994, the city's fire department began working with NASA Langley to improve firefighter equipment. According to the Mid-Atlantic Technology Applications Center (MTAC), which coordinated the agreement between the agencies, "Innovations being investigated include an advanced tracking system for locating personnel and equipment in a burning structure; noise reduction and control technologies to improve communications with the on-scene command post; in-mask carbon monoxide management through application of Langley's low-temperature oxidation catalyst, coupled with a warning system; and new lightweight, heat-resistant, and self-extinguishing materials for use in equipment, clothing, and hoses."

NASA has been instrumental in several efforts to improve the safety and comfort of firefighters. Kennedy Space Center (KSC) developed a liquid air pack for Space Shuttle astronaut rescue crews and for use during hazardous operations at the center, according to a press release issued by the center in 1996. "It is lighter by several pounds and more compact than conventional air packs to allow rescue crews to enter the narrow passages within the orbiter in the case of an emergency at the launch pad." Plus, it provides at least twice as much breathing time as standard air packs.

The liquid air packs combine liquid oxygen and liquid nitrogen, which evaporate as they travel from the pack to the facemask. The gases are a cool 65 to 70 degrees Fahrenheit, helping cool down the firefighter's core body temperature.

The concept is similar to the SCAMP—which stands for Supercritical Air Mobility Pack—made by Aerospace Design & Development, Inc. The Colorado company was asked to produce a breathing apparatus that would allow Kennedy Space Center rescue personnel to crawl through a 20-inch opening—a requirement for a launch pad emergency. The result of their research is a self-contained breathing apparatus with a high-density capacity air pack that uses cryogenic (very cold, in this case about –320 degrees Fahrenheit) air and can work in microgravity environments. On Earth, that means it can work regardless of the position of the supply tank. But unlike other liquid, air breathing system users don't suffer oxygen enrichment. A one-hour breathing/cooling system weighs approximately 25 pounds; the two-hour system, about 35 pounds.

The SCAMP pack is shaped like a thin flat box, allowing firefighters to get through narrow spaces.

See also:

Digital Nose, Infrared Detector Improvements

Additional Reading

"Electronic Nose" *NASA Spinoff 2001* (2001): 72.

"Emergency Response Breathing Apparatus" *NASA Spinoff 2000* (2000): 44.

"Fire Prevention in Space" *NASA Explores* 27 September 2001, http://www.nasaexplores.com/show2_article.php?id=01-064.

"Fire Resistant Materials" *NASA Spinoff 1982* (1982): 98.

"KSC Liquid Air Mixer Key to Improved Firefighter Air Pack." Kennedy Space Center press release, 1 October 1996.

"NASA Anti-Fire Device Protects Firefighters." *Innovation* 5, no. 1 (January/February 1997).

"NASA Brings Spacesuit Technology to Fire Fighting." *SAMPE Journal*, July/August 2001.

"Space Technology for the Fire Department" *NASA Spinoff 1986* (1986): 50.

"Spacesuit Technology May Aid Firefighters." *Aerospace Technology Innovation* 8, no. 6 (November/December 2000).

"Spacesuit-based Firefighter Suit to Be Shown at I-2000" Johnson Space Center press release, 4 October 2000.

Food Preparation

Just as Mercury astronauts proved to the world that human space flight was possible, they also proved that it was physiologically possible to eat and drink in the microgravity of a spacecraft. The concern wasn't so much what they ate—because the flights were very short. Their meals could be eaten before liftoff. During the flight, however, they experimented by eating bite-sized cubes of food, freeze-dried foods, and foods packed in aluminum tubes.

The aluminum tubes were often heavy and awkward to handle. Freeze-dried foods were rehydrated in the astronaut's mouth by his saliva. To ensure that cubes of food didn't crumble—because the crumbs could lodge in sensitive equipment—the food was coated with gelatin. The flights showed that these food preparation processes worked, but the food wasn't very tasty.

Since then, food preparation has come a long way and the technology to provide better tasting and more nutritious foods to astronauts has had a great deal of impact on how foods are prepared and packaged here on Earth.

The first big improvements came during the Gemini missions. Scientists freeze-dried the foods. According to NASA, "Freeze-drying techniques in the space program consist of slicing, dicing, or liquefying prepared food to reduce preparation time. After the food has been cooked or processed, it is quick-frozen, then placed on drying trays and put into a vacuum chamber where the air pressure is reduced. Heat is then applied through heating plates. Under these conditions of reduced pressure and increased temperature, the ice crystals in the frozen food boil off, and the water vapor that is left is condensed back to ice on cold plates in the vacuum chamber. Because water is the only thing removed in this process, the freeze-dried food has all the essential oils and flavors. The texture is porous and can be easily rehydrated with water for eating.

"To rehydrate food, water was injected into the package through the nozzle of a water gun. The other end of the package had an opening in which the food could be squeezed out of the package into the astronaut's mouth. Because of the size of the opening, food particle size was limited. After the meal had been completed, germicidal tablets were placed inside the empty package to inhibit microbial growth on any leftovers."

Gemini astronauts repeated a menu every four days, but the variety was far greater than during Mercury missions. Gemini astronauts sampled grape and orange drinks, cinnamon toasted bread cubes, fruit cocktail, turkey bites, applesauce, cream of chicken soup, shrimp cocktail, beef stew, chicken and rice, and turkey and gravy, according to NASA sources.

Two developments occurred during the Apollo program that made eating in space seem more like eating on Earth. NASA developed a spoon bowl—a plastic container into which water was added to rehydrate food. Because the food was moist after it had been rehydrated, it would cling to the spoon. That meant astronauts no longer needed to drink through straws or package openings. The other new package was called the wetpack—a flexible pouch with aluminum foil laminate or a can. The canned foods weighed about four times more than freeze-dried foods, which was their main disadvantage. But foods packaged in wetpacks looked and smelled like meals on Earth. Apollo mission menus included coffee, corn flakes, scrambled eggs, beef sandwiches, tuna salad, beef pot roast, spaghetti, frankfurters, chocolate pudding and tuna salad.

On *Skylab*, mealtime became even more like its Earthbound counterpart. *Skylab* had a freezer, refrigerator, warming trays and a table, according to NASA reports. "The supply of food on board was sufficient to feed three astronauts for approximately 112 days... *Skylab* foods were packaged in specialized containers. The rehydratable beverages were packaged in a collapsible accordion-like beverage dispenser. All other foods were packaged in aluminum cans of various sizes or rehydratable packages," NASA explains on the portion of its Web site dedicated to space foods. *Skylab* astronauts could select their food packages and put them in a conduction-warming tray.

By the time the Space Shuttle program began, there were 74 different kinds of food and 20 choices of drinks for astronauts. The shuttle has a galley with water dispenser and oven. Hot, cold or room-temperature water can be used and food is warmed in its own package. NASA sources say it takes about four minutes for a full meal for a crew of four to be set up, although reconstituting and heating food takes about a half hour more.

Meal trays double as dinner plates by attaching to the astronauts lap or to a wall. Astronauts use common utensils, plus scissors to open their food packages.

More wetpack food than rehydratable food will be available on the International Space Station because water will not be as readily available as it has been on Shuttle missions.

Some of the commercial spin-offs for food preparation techniques are fairly obvious. Backpackers on long hikes pack freeze-dried food because it is of much lighter weight than conventional food. Oregon Freeze Dry, Inc. supplies much of the freeze-dried food for Shuttle flights. The company worked with NASA in the 1960s to adapt its freeze-drying technology to NASA specifications. Its Mountain House brand of freeze-dried products for backpackers is very similar to space food. The company also sells some freeze-dried space foods, including space ice cream, on its Web site.

In the mid-1970s, NASA was involved in a demonstration project in which packaged meals were delivered to elderly people in need. The goal was to provide a greater variety of nutritious foods to these particular people who were not involved in another food delivery program. The packages were either delivered in person or were mailed to participants. Based on surveys conducted after the project ended, most people said they liked the food, the quantities, and they liked the ease with which it was prepared. Canned food was generally preferred over freeze-dried food, but most people were willing to eat freeze-dried meals occasionally. According to a newsletter published by the Lyndon B. Johnson School of Public Affairs, "Some 40 percent indicated that their eating habits changed while on the program. Of this number, most reported they were eating both a greater amount and a greater variety of food than before."

Distributing foods using NASA technology is one straightforward spin-off. However, much of the technology that went into making food and mealtimes safe on board spacecraft has also been incorporated into packaging. For example, the Pillsbury Co., of Minneapolis, Minnesota, spent more than a decade working with NASA to produce some of the first space foods, according to a NASA article, "NASA Space Food Technology Brings Safe, Tasty, and Nutritionally Balanced Meals to Earth." Pillsbury solved the problem of crumbling food by coating bite-sized foods with an edible gelatin coating. The bigger problem was to develop a system to avoid bacterial contamination.

Pillsbury developed the Hazard Analysis and Critical Control Point (HACCP) concept. The NASA article explains: "The first step, hazard analysis, is a systematic study of a product, its ingredients, processing conditions, handling, storage, packaging, distribution, and directions for consumer use to identify sensitive areas that might prove hazardous. Hazard analysis provides a basis for blueprinting the Critical Control points (CCPs) to be monitored. CCPs are points in the chain from raw materials to finished product where loss of control could result in unacceptable food safety risks."

For example, the article details those CCPs for cooking and packaging turkey breast. Each point in the process has certain safety requirements, such as the internal temperature. The result, the article explains, is "a sophisticated process control system highly unlikely to produce an unsafe or otherwise contaminated product."

Not only was the HACCP system used in packaging space food, Pillsbury plants had begun using the system for all its products. "Pillsbury's subsequent training courses for Food and Drug Administration (FDA) personnel led to the incorporation of HACCP in the FDA's Low Acid Canned Foods Regulations, set down in the mid 70s to ensure the safety of all canned food products in the U.S."

The development of the HACCP system has truly become the cornerstone of food safety in this country. The FDA has added HACCP for seafood and juice products; the U.S. Department of Agriculture has added it for meat and poultry processing and packaging.

While food processing for space flight has come light years from its start in aluminum squeeze tubes, NASA still has years of research ahead of it. Soon, NASA will be testing the life support systems in its Bioregenerative Planetary Life Support Systems Test Complex (BIO-Plex). These systems will be used in the Mars mission scheduled for 2014. A trip to Mars is expected to take six months and an 18-month stay on the planet is anticipated. That means the Mars crew would be gone at least two and a half years, requiring even more planning and preparation for safe, nutritious, and fresh foods. And if foods are actually prepared in space, the life support systems in place will have to be able to handle any hazardous compounds created by cooking or processing.

Additional Reading

"A Dividend in Food Safety" *NASA Spinoff 1991* (1991): 52.

"Eating Right for Long-Duration Space Missions." Johnson Space Center press release, 2 July 2001.

"Food for Space Flight" NASA Fact Sheet, http://www.jsc.nasa.gov/pao/factsheets/food.pdf. Accessed 8 June 2003.

Formanek, Raymond Jr. "Food for Thought: An Interview with Joseph A. Levitt." *FDA Consumer* 35, no. 5 (September 2001): 12.

Friske, Marcus and Annie Platoff, compilers. "Meal System and Food Processing: NASA Space Food Technology Brings Safe, Tasty, and Nutritionally Balanced Meals to Earth." 15 May 1996 http://www.jsc.nasa.gov/pao/spinoffs/mealsys.html. Accessed 20 April 2002.

Hollingsworth, Pierce. "Federal R&D: A Boon to Food Companies." *Food Technology* 55, no. 6 (June 2001): 45.

"Meals for the Elderly." *NASA Spinoff 1980* (1980): 116.

Oregon Freeze Dry Inc Web site. www.mountainhouse.com/data/space-fd.html.

"Packaged Food" *NASA Spinoff 1976* (1976): 87.

Purvis, Hoyt H., ed. "Meals for the Elderly Project Reports" *The Record* (Lyndon B. Johnson School of Pubic Affairs, The University of Texas at Austin) no. 24 (17 May 1976).

"Space Food" *NASA Spinoff 1994* (1994): 82.

"Space Food and Nutrition Educator's Guide." A comprehensive report on technologies involved in space food development, menu items and classroom activities. http://spacelink.nasa.gov/Instructional.Materials/NASA.Educational.Products/Space.Food.and.Nutrition. Accessed 19 April 2002.

Fresher Vegetables

The first humans in space didn't stay there long enough to get very hungry. On later, slightly longer missions, it was easy enough to bring prepared foods with them. But on future missions, when manned spacecraft make journeys requiring years in space, they'll need more food than they can carry.

So NASA researchers developed a fully enclosed greenhouse, called the Astroculture™ system. It has been used on Space Shuttle missions since 1993. Today, the technology is helping grocers and horticulturalists who use it to keep produce and flowers fresh longer.

Creating a space-based greenhouse is a difficult task for a number of reasons. Growing plants in a sealed environment requires balancing nutrients and light and removing the naturally produced gases plants emit that could build up to unhealthy levels and could kill the crops.

The system was used for experiments aboard Russia's *Mir* Space Station, where it showed that seeds from plants grown in space could be planted and new seeds harvested, proving that it would be possible for humans to grow their own foods and to survive long periods in space. On previous Space Shuttle missions it's been used to grow potatoes, wheat, mustard plants and even a miniature rose. And it will be used on the International Space Station, which began taking shape 220 miles above the Earth when its first component was launched in 1998.

The system to be used on the space station isn't big—it can only handle plants up to 15 inches high. Its first experiment will be to study whether *Arabidopsis*, of the *Brassica* plant family, can complete its seed-to-seed life cycle in microgravity. Scientists are still learning about the long-term effects of weightlessness on plants and animals and learning how best to make them thrive in space.

This space greenhouse was created at the Wisconsin Center for Space Automation and Robotics (WCSAR), a NASA Commercial Space Center sponsored by the Space Product Development Office at NASA's Marshall Space Flight Center. What makes it unique is that researchers figured out a way to remove ethylene gas. Ethylene is a natural hormone that is produced by plants as they ripen. Too much of it causes plants to wither and spoil. "You've probably brought a tomato home and it looked pretty good the day that you brought it home," University of Wisconsin-Madison Professor Marc Anderson said in a 1998 interview with CNN.com. "But the next day it's got this little spot and you open it up and it's not very good, and that's typical of ethylene damage."

Anderson led the team that in the mid-1990s discovered that titanium dioxide (a harmless coloring agent used in many consumer products) exposed to ultraviolet light would convert the ethylene to carbon dioxide and water—both good for plants. Today that solution is helping grocers and florists, who face the same problems of wilting plants. An Atlanta-based company, KES Science and Technology, uses the same technology to make devices that can be installed in refrigerated display cases at the grocery store or flower shop, or in refrigerated trucks. The company's box-shaped Bio-KES devices are "ethylene scrubbers." They come in a range of sizes—small enough for a flower-shop case and big enough for a tractor-trailer. And they can save a lot of money and reduce the waste of throwing out rotten food and flowers.

Peaches, for example, are highly sensitive to ethylene gas. Without ethylene removal a peach could last just seven days after being harvested. By cutting the ethylene exposure and keeping the peach in humid air, it could last up to 28 days, the company says. Rotting produce is expensive: high levels of ethylene gas accounts for up to 10 percent of produce losses and 5 percent of flower losses, according to KES.

Another feature of the technology is that it is almost maintenance free. Previous systems required frequent maintenance to replace the oxidant, usually potassium permanganate, which was being used to remove the ethylene gas. The glass pellets coated with titanium dioxide never need replacing so long as they're not damaged. The carbon dioxide and water the system produces can simply be piped back into the greenhouse or storage area.

This new produce-preserving technology has a lot of potential. It could reduce the resources used to grow the world's food supply and help make the food we already grow reach the consumers before it decomposes. KES President John Hayman,

Jr. told CNN.com, "I couldn't even begin to give you the billions of tons of food that can be given, or handed to the end user, in a usable state, rather than something that the grocery stores or the commercial enterprises have to throw out."

Additional Reading

"Advanced Astroculture." *NASA Fact Sheets* (FS-2001-03-47-MSFC), Marshall Space Flight Center Web site. http://www1.msfc.nasa.gov/NEWSROOM/background/facts/advasc.html. Accessed 19 April 2002.

Frazer, Lance. "Titanium Dioxide: Environmental White Knight?" *Environmental Health Perspectives* 109 (April 2001): A174.

"Fresh Veggies from Space." *NASA Spinoff 2001* (2001): 80.

KES Science & Technology, Inc. Web site. www.storagecontrol.com/kes.htm.

Lockridge, Rick. "Shuttle Experiment Leads to Longer-Lasting Produce." CNN.com. 17 September 1998. http://www.cnn.com/TECH/science/9809/17/t_t/produce.protector/index.html. Accessed 27 June 2003.

Glasses and Sunglasses Improvements

Thanks to NASA research, eyeglasses and sunglasses today are more resistant to scratching and are better able to absorb harmful ultraviolet rays. In the first example, NASA technology to improve the scratch-resistance of plastics was directly applicable to spectacle lenses. The second example, however, is a second-generation spin-off—NASA technology was first applied to welding curtains and then that technology was applied to spectacle lenses.

Prior to the early 1970s, most eyeglass lenses were made of glass. Glass could be ground to provide excellent optics. But, of course, glass breaks. When the Food and Drug Administration ruled in 1972 that sunglasses and spectacles needed to be shatter resistant, the optical industry turned to plastic. Plastic provides nearly the same visual quality, doesn't shatter and weighs much less. The biggest disadvantage to plastic is that it scratches easily.

Scratch-resistant sunglasses. Photo courtesy NASA.

Researchers at NASA Lewis Research Center created a film of diamond-like carbon (DLC) to be applied to the lenses. The material mimics the hardness of diamonds. According to NASA, "DLC employs the advantages of diamonds without the cost. A thin layer of DLC is deposited on the lens by using an ion generator to create a stream of ions from a hydrocarbon gas source; the carbon ions meld directly on the target substrate and 'grow' into a thin DLC film. The coating offers 10 times the scratch resistance of conventional glass lenses."

The coating was intended to protect plastic surfaces of aerospace equipment, including windows and face shields. It was brought into the commercial market by sunglass maker Foster Grant. In 1991, a new subsidiary Fosta-Tek (Leominster, MA) assumed the license for the NASA process. The company uses it on eyewear, face shields, and gas masks for military and industrial uses.

Another NASA spin-off related to eyewear is ultraviolet radiation-blocking lenses. Research has shown that ultraviolet radiation could be a contributing factor in certain eye-related diseases, such as the development of cataracts, cancer of the skin around the eye, and it may contribute to age-related macular degeneration. Astronauts in space are exposed to higher amounts of UV radiation.

In the 1970s, the welding industry came to NASA's Jet Propulsion Laboratory (JPL) for help. The industry wanted welding curtains that would protect welders from blue and ultraviolet radiation while they were working. But for safety's sake, the curtain needed to be see-through. According to NASA, two JPL scientists spent three years applying their problem-solving techniques to the development of a dye formula for a welding curtain. A decade later, that same concept was spun off into high-performance sunglasses that provide maximum UV protection. They help aviators, sailors and others see more clearly in bright light, haze, and fog.

The company that makes these sunglasses, SunTiger, explains its NASA connection on their Web site.

> Almost two decades ago, NASA scientists began researching the superior vision of eagles, falcons, hawks and other birds of prey. They discovered that these birds secrete an orange colored retinal fluid that absorbs harmful ultraviolet, and shorter blue light wavelengths, giving the bird's eyes protection from bright sunlight while enhancing contrast.
>
> Through this research scientists developed an organic dye that has beneficial effects for human vision when incorporated into a sunglass lens. This revolutionary technology led to the development of the SunTiger and Eagle Eyes lenses.

The lenses block 99 percent of the potentially harmful wavelengths.

NASA filters have also been used for other purposes. For example, a low-cost brown filter that was originally developed at NASA's Ames Research Center to help farmers identify diseased plants was modified by Optical Sales Corp. (Portland, OR), in 1997 to help improve vision for pilots and drivers. The brown filter blocks much of the yellow and green light during daylight hours, so the eye can better detect the other colors in the visible spectrum. Developer Leonard Haslim explained the original intent of the filters in a 1997 NASA press release "Stress in plants tends to be camouflaged by the plant's natural chlorophyll. As a result, many plant diseases cause irreversible damage by the time they become visibly evident. In the past, it was necessary to have highly trained professionals examine plants in order to determine signs of stress in the early stages. . . . 'Now, farmers themselves can use goggles equipped with the special filter to locate diseased or stressed plants. Sick leaves that appear just a bit yellow in normal light show up as a much brighter yellow when viewed through the filter. Conversely, healthy leaves appear as a vivid green,'" Haslim said.

See also:

Anti-Fog Spray, for another example of better functioning of glasses, goggles and masks

Additional Reading

"Commercialization of NASA Filter to Aid Pilots/Drivers." NASA Press release, 17 September 1997.

Fosta-tek Web site. www.fostatek.com.

"Radiation Blocking Lenses" *NASA Spinoffs: 30 Year Commemorative Edition* (1992): 56.

"Scratch-Resistant Sunglass Coating" *NASA Spinoffs: 30 Year Commemorative Edition* (1992): 84.

SunTiger Web site. www.suntiger.com.

Golf Ball Improvements

The connection between golf and the moon goes all the way back to the *Apollo 14* mission in 1971. As the second moonwalk was coming to an end, astronaut Alan Shepard dropped two golf balls on the moon's powdery surface and hit them with a makeshift golf club. In an interview with Johnson Space Center in February 1998, he said, "Being a golfer, I was intrigued before the flight by the fact that a ball with the same club head speed will go six times as far. Its time of flight—I won't say 'stay in the air,' because there's no atmosphere—will be at least six times as long. It will not curve, because there's no atmosphere to make it slice."

Years later, engineers were still using research developed for space travel to figure out how to make golf balls soar better here on Earth. For example, Wilson Sporting Goods Company incorporated a design that uses NASA aerodynamics technology to create its Ultra (TM) 500 Series golf ball in 1995. The 500 dimples on the ball are arranged in a pattern of 60 spherical triangles of different sizes, shapes, and depth. Earlier golf balls had about 20 triangular faces.

Wilson Golf assigned Robert T. "Bob" Thurman to the project. Thurman had earlier been an engineer with Marietta Manned Space Systems, the manufacturer of the Space Shuttle's external tank. What the research showed was that the size and placement of these patterns made a difference in how well the golf ball soared through the air. In the NASA publication *Spinoff 1995*, Thurman said, "In a way, my job is very similar to that at Martin Marietta. Instead of analyzing airloads on the Shuttle, I analyze airloads on airborne golf balls and the effects they have on the ball's trajectory, shape and distance. Instead of analyzing the slosh damping capability of the Shuttle's liquid oxygen tank, I analyze the effects of varying moments of inertia on the spin decay of a golf ball due to its liquid center."

Testing showed that large dimples reduce air drag, enhance lift, and maintain spin for distance. Small dimples prevent excessive lift that would destabilize the ball's flight. Medium dimples combined some of the characteristics of both. By creating a formula for placement of the ball, Wilson developed a "more uniform airflow over the spinning golf ball surface."

NASA technology also helped the Ben Hogan Company researchers analyze different designs for new golf balls. High-speed video equipment was developed at the NASA Glenn Research Center in Ohio. Rather than taking regular images of aircraft engines, this video equipment can record thousands of images per minute. As a result, scientists can analyze smaller details. The equipment was used to obtain accurate information for lunar missions and in aircraft engine development. Ben Hogan Company—now part of Spalding—officials contacted the center to see if the high-speed imaging system could help them evaluate the spin rates of their different golf ball designs.

Additional Reading

"Get the Jump on Sports Equipment." *Materials Engineering* 109, no. 6 (June 1992): 10.

"Golf Aerodynamics." *NASA Spinoff 1995* (1995): 76.

Johnson Space Center Web site. Interview with Alan Shepard. 20 February 1998. http://www.jsc.nasa.gov/pao/shepard/transcript.html. Accessed 6 April 2002.

Rovito, Rich. "Contract Will Bring NASA Technologies to Wisconsin Manufacturers." *The Business Journal—Milwaukee* 18, no. 33 (4 May 2001): 6.

Skorupa, Joe. "The Aerospace Connection." *Sporting Goods Business* 29, no. 2 (February 1996): 108.

Stogel, Chuck. "Techno-legit at Wilson." *Brandweek* 36, no. 16 (17 April 1996): 18.

Golf Club Improvements

Strong, durable and malleable, metal in its many forms—such as aluminum, steel, magnesium, iron and gold—has proven to be one of the most remarkable materials ever discovered. It's hard to imagine what our everyday life would be like without it.

Space exploration can often require even more advanced materials. So scientists are constantly experimenting with new metal alloys. In 1992, researchers working on project sponsored by NASA's Jet Propulsion Laboratory in Pasadena, Calif., achieved a breakthrough discovery. In their search for a new aerospace material, Professor William L. Johnson and Dr. Atakan Peker discovered a new kind material that is stronger—and lighter—than steel or titanium.

The material is called Vitreloy, and it is more like a glass than a metal. Comprised of zirconium, beryllium, copper, titanium and nickel, it is in a class of materials called vitrified metals, also known as metallic glass. The difference between Vitreloy and regular metal is in its atomic structure. Conventional metals have a crystalline structure in which the atoms align themselves in recurring grain patterns. In Vitreloy, the atoms don't organize in any pattern—it is more like a liquid than a metal. And, the absence of grain patterns means there are no weak points where fractures can occur. Vitreloy, according to NASA, is twice as strong as titanium or steel, but softer and more malleable.

The Cal Tech researchers who developed it were thinking about aerospace uses such as jet wings or rocket-engine parts. But one of the researchers, Bill Johnson, was an avid golfer who saw another potential, according to a March 1998 *Sports Illustrated* article. Johnson told investors at Amorphous Technology International (ATI) of Laguna Niguel, California. Soon after, the company licensed the material from Cal Tech and began using it to make a better kind of golf club.

Vitreloy—or Liquidmetal, as it is called by the golf-club makers—has a couple of characteristics that help golfers. First, its high strength and low weight means it absorbs less of the energy from the impact with a golf ball. So, more energy is transferred directly to the ball, making it go farther. Titanium-head clubs transfer about 70 percent of the energy to the ball, while Liquidmetal clubs transfer 99 percent, according to Howmet Metal, the Michigan company that handles production of the clubs for ATI. Golfers report that clubs made with Liquidmetal have a softer, more solid feel and a bigger "sweet spot," which makes them very forgiving for golfers who have trouble hitting the ball straight.

The company is looking for ways to use Vitreloy in other recreation products such as tennis rackets and baseball bats. Because it can be made in a highly bio-

Liquidmetal golf clubs. Photo courtesy NASA.

compatible form, Vitreloy has potential uses in medical components such as prosthetic implants and surgical instruments, according to NASA.

The breakthrough that helped scientists create Vitreloy is already helping space exploration. The Genesis spacecraft, which was launched in August 2001 on a three-year mission, uses a small, disk-shaped piece of metallic glass to collect ion samples of solar wind a million miles from Earth. When the spacecraft returns to Earth in 2004, the samples it has collected will help scientists study the interstellar cloud of gas and dust known as the solar nebula, which was behind the creation of our solar system billions of years ago.

This newly developed substance is being used to gather ion samples because it absorbs and retains helium and neon, which are important to understanding solar and planetary processes, according to NASA. When melted, the surface of the metallic glass dissolves evenly, letting scientists control the release of the captured ions in the laboratory. Engineers for NASA are also using it to build a drill to let astronauts search for water beneath the surface of Mars.

Golf has benefited from other space-related technologies as well. Engineers working at the Marshall Space Flight Center on the design of the International Space Station developed a metal alloy dubbed "memory metal" for its ability to return to its original shape if deformed. Memry Corp. of Brookfield, Connecticut, which worked with the Marshall Center in developing alloys for the space station, then developed the technology into an alloy used in golf clubs made by Nicklaus Golf Equipment of West Palm Beach,

Florida. When the club strikes the ball, the alloy's elasticity lets the ball stay on the club face a split-second longer—that puts more spin on the ball and gives the golfer greater control. Plus, NASA is continuing its research into what it calls "smart" materials. At the Langley Research Center memory metals are part of an effort to develop aircraft with flexible wings that would be able to unfurl on command.

Additional Reading

Ashley, Steven. "Metallic Glasses Bulk Up." *Mechanical Engineering-CIME,* June 1998, 72.

Barry, Patrick L. "Buck Rogers, Watch Out!" *NASA Science News.* 1 March 2001. http://science.nasa.gov/headlines/y2001/ast01mar_1.htm. Accessed 8 June 2003.

Folkers, Richard. "Titanium Tee Time." *U.S. News & World Report* 125 (13 July 1998): 59.

Larsen, Don. "Liquidmetal Swings into Action." *Materials World* 6, no. 9 (September 1998): 547.

Lipsey, Rick. "Betting on Big Ideas." *Sports Illustrated,* 16 March 1998, 40.

"Memory Metal Technology Finding Its Way Into Many Products." NASA Southeast Technology Transfer Alliance, bulletin 2, no. 2. Summer 1997. http://technology.ksc.nasa.gov/TECHNO/alliance_bulletin8.html. Accessed 8 June 2003.

Proctor, Paul. "AKA Liquid Metal." *Aviation Week & Space Technology* 148, no. 12 (23 March 1998): 17.

"Through a Metal, Darkly." *The Economist,* 13 March 1999, 95.

Hang Gliders

Decades before hang gliding became a popular sport, NASA gave the concept a big push forward. Parawings or hang gliders were originally developed in 1948 for use as a wing on inexpensive aircraft. Over the next decade, NASA conducted research into parawings as a way to bring space payloads back to Earth.

In July 1961, astronaut Gus Grissom splashed down in his Mercury capsule called Liberty Bell. The capsule had a newly designed escape hatch on it, which Grissom would open when rescue helicopters came to the scene of the splashdown. While waiting, however, the hatch accidentally blew open and seawater started flooding the capsule. The capsule sank, and nearly took Grissom with it. This incident was the most serious in a series of difficult recoveries of Mercury capsules. Instead of using unguided parachutes for the upcoming Gemini and Apollo capsules, the agency wanted more controllable recovery systems.

One system with which the agency experimented was the Rogallo wing, a flexible wing pioneered by Francis Rogallo, a NASA engineer. During reentry, the astronauts could deploy the Rogallo wing like a parachute, and would be able to direct the capsule to a touchdown spot. The wing would also create drag and slow the capsule down.

According to the National Air and Space Museum's *Aircraft of the Smithsonian,* NASA began testing the Rogallo wing in 1962 on its Paresev research vehicle. "During early test flights, automobiles towed the Paresev across the Edwards Dry Lake test range in California. Later, aircraft towed the Paresev to altitude and glide tests were conducted. NASA eventually discarded the Rogallo recovery method in favor of simpler, more reliable and more economical parachutes, but publicity from the Paresev test sparked interest in the design among several tinkerers."

Among these tinkerers were Australians John Dickenson and Bill Bennett, who moved to California, and by 1969 they were producing these kite wings under the

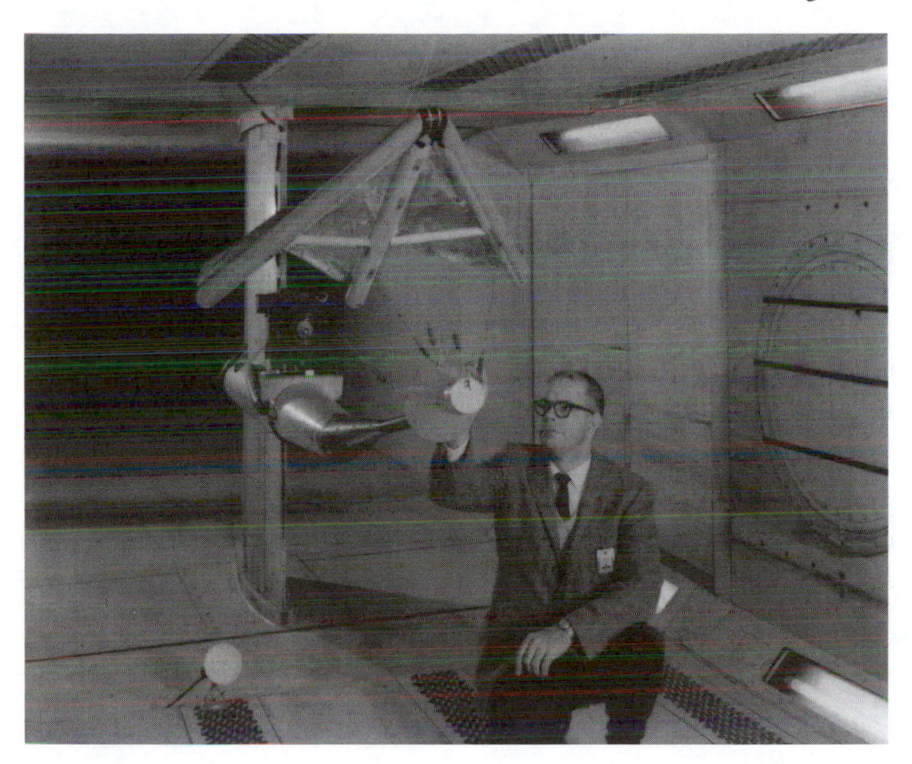

Francis Rogallo with a model of the paraglider, or "Rogallo Wing." Photo courtesy NASA.

Delta Wing name. It was a lightweight and durable triangle-shaped airfoil. To launch, a hang glider pilot runs down a slope to get air moving across the wing. That generates lift. While gravity pulls the pilot and wing back down, the opposing forces move the hang glider forward.

The National Air and Space Museum history of hang gliding calls Bennett "a spectacular promoter of the infant sport," cataloging his flights near the Statue of Liberty on the 4th of July, 1969, and his stint in 1972 as a stunt double for actor Roger Moore playing James Bond in the movie *Live and Let Die*.

The sport went to new heights—literally—when pilots frustrated with often-risky ways of towing their hang gliders into the air began seeking out seaside cliffs and mountainsides as launching pads. According to the Smithsonian history, "Much of this activity happened in California, where the counterculture movement happily embraced it. Frequent crashes made the learning curve steep and painful, but when 1973 ended, Bennett and others were successfully gliding and even soaring their Rogallo wings using thermal updrafts and ridge lift." Thermal updrafts are columns of hot air that provide lift and ridge lift is air deflected upward by mountainous terrain.

In Kitty Hawk, North Carolina—where the Wright Brothers took their first flight—Francis Rogallo and his wife Gertrude are celebrated as pioneers. Gertrude Rogallo helped to make the first Rogallo wings out of dining room curtains in the family's kitchen. Rogallo turned 90 in January 2002 and the occasion was marked by a retrospective in an online newsletter.

Additional Reading

Connor, R. and R. E. Lee. "Aircraft of the Smithsonian" (revised). National Air and Space Museum, Smithsonian Institution Web site. 14 September 2001. http://www.nasm.si.edu/nasm/aero/aircraft/delta_wing_162.htm. Accessed 8 June 2003.

"Hang Gliders" *NASA Spinoff 1981* (1981): 79.

Lowry, John G., Delwin R. Groom, and Robert T. Taylor. *Preliminary Investigation of a Paraglider.* NASA Technical Note D-443. Langley, Va., 1960.

"PARESEV: A Space-Age Kite" *On the Frontier: Flight Research at Dryden 1946–1981* (chapter 7-4). http://www.dfrc.nasa.gov/History/Publications/SP-4303/ch7–4.html. Accessed 8 June 2003.

Rogallo, Francis M. "Parawings for Astronautics." *Advances in the Astronautical Sciences* 16, part 2 (1963): 3–7.

Swedrock, Karen. "Father Of Hang Gliding Turns 90." February 2002. North Carolina's Outer Banks Web site. http://www.outer-banks.com/ecnews/index.asp#feb06. Accessed 27 March 2002.

Healing Light

Light is a powerful tool. For Space Shuttle plant-growth experiments, scientists from NASA's Marshall Space Flight Center in the mid-1990s had developed light-emitting diodes (LEDs) that could accelerate the growth of food. The experiments were part of an effort to develop ways to grow food in the confines of a space ship, and the experiment was part of a 1995 Space Shuttle mission. Scientists used the light of LEDs, which have a longer wavelength and are near infrared, because it increases the energy produced in the mitochondria of plant cells, which makes the cells grow faster.

That same power of light is also being used to treat cancer. Called "photodynamic therapy," the procedure involves injecting photosensitive, tumor-killing drugs into cancer patients and then using light from lasers to active the drugs. While the treatment was effective, it also presented some problems. Lasers were expensive and sometimes unreliable, could only treat smaller areas, and had a tendency to overheat the tissue surrounding the treatment area.

That's where the LEDs developed for NASA proved effective. The researcher who had developed the LEDs for plant-growth experiments approached doctors with the idea of substituting LEDs for lasers in cancer treatments. In a series of experiments beginning in the mid-1990s, Dr. Harry Whelan, a professor of pediatric neurology at the Medical College of Wisconsin, used the NASA-developed LEDs to treat solid-tumor cancers. The treatment involved injecting patients with a light-sensitive drug. The drug permeates the tumor, and is activated by the light from an LED probe. Once activated it generates molecules known as free radicals, which destroy the unwanted tissue and leave the surrounding tissue undamaged.

The LED probe used by Dr. Whelan was developed by Quantum Devices, the same Wisconsin company that had developed the LEDs for the Space Shuttle plant-growth experiments. The new probe consists of a hollow steel tube that holds 144 tiny diodes, and is only 10 centimeters long. In an interview with Medical College of Wisconsin publication *HealthLink*, Dr. Whelan described the advantages of the LED device: Because it produces a longer (red) wavelength, "it permeates deeper into tissue, and allows us to give new hope to patients whose tumors previously were too deep to be reached." While a laser focuses its light on

LED wound-healing device. Photo courtesy NASA.

a single point, LEDs cover a much larger area, and thus can be used to treat the entire body. Also, LEDs operate at a cooler temperature than lasers and can be used safely for repeat treatments. On face, the device can be used for hours and still not become too hot to touch. Currently, photodynamic therapy using a drug called Photofrin is a standard therapy for some types of lung cancers and advanced esophageal cancer.

LED therapy alone produces cell growth and tissue regeneration by reacting with cytochromes in the body. Cytochromes are the parts of cells that respond to light and color, and when activated they produce the energy that leads to tissue growth. LED therapy has been used to successfully treat wounds such as diabetic skin ulcers and burns.

The healing power of LED therapy as used on Earth has applications for space travel. In microgravity, wounds heal more slowly, and muscles and bones are prone to atrophy. To treat wounds, researchers developed a small (3.5-inch by 4.5-inch), flat array of LEDs. Dr. Whelan said in an interview with NASA: "Using an LED array to cover an astronaut may help prevent the effects of microgravity. LED therapy could also be used to help treat conditions that could arise in space that don't respond to treatment because of those microgravity situations. A simple cut might heal faster with LED, but the benefits would be even more notable if an astronaut suffered a severe injury." According to Dr. Whelan the technology can be used in different applications. The devices for LED therapy in space were designed to be light, which allows the military to treat wounded soldiers in the field. The devices also can be used under the sea. The USS *Salt Lake City*, a Navy submarine, carried an LED device while deployed in the Pacific Ocean and reported that crewmembers' cuts healed 50 percent faster using LED therapy.

Additional Reading

Goodell, Teresa Tarnowski and Paul J. Muller. "Photodynamic Therapy: A Novel Treatment for Primary Brain Malignancy." *Journal of Neuroscience Nursing* 33, no. 6 (December 2001): 296.

"Killing Cancer with Light." MWC HealthLink. Web site. 21 January 1999. http://healthlink.mcw.edu/article/916936587.html. Accessed 8 June 2003.

"LEDs Developed by NASA Used to Ablate Brain Tumors." *Oncology News International* 9, no. 1 (January 2000).

Medical College of Wisconsin's Light-Emitting Diode Web site page. www.mcw.edu/whelan.

"NASA Aids Fight Against Cancer." *NASA Science Information Systems Newsletter* 45 (1997).

"NASA Light Emitting Diode Medical Applications from Deep Space to Deep Sea." Presented at the Space Technology and Applications International Forum-2001 by Dr. Harry T. Whalen et al., 11–14 February 2001.

Quantum Devices, Inc. Web site. www.quantumdev.com.

"Space Research Shines Life-Saving Light." *Aerospace Technology Innovation* 5, no. 6 (November/December 1997).

Heart Monitor Improvements

When the first astronauts were getting ready to go into space, nobody was really sure how their bodies would respond. In many ways the workings of the digestive system, the heart, and other organs are affected by the pull of gravity. Some physicians involved in the early days of the space program worried that the body's organs couldn't function properly without gravity. According to a NASA account of the Mercury mission planning, physicians worried that weightlessness would make it impossible for the astronauts' brains to control their bodies and leave them "with an absolute incapacity to act."

The heart and circulatory system were especially critical, since a disruption in the flow of blood could cause astronauts to become dizzy or event faint. And with every new mission putting astronauts into space for longer periods, scientists were searching for ways to measure the body's response in real time instead of waiting for astronauts to return to Earth.

In the mid-1960s, scientists working for NASA developed a practical way to tell them exactly what happened to an astronaut's heart during space flight. The technology is called impedance cardiography. It sends a low-level electrical impulse into the body and measures it after it passes through the heart, using electrodes attached to the chest and neck. The impedance to the current shows doctors how much blood is being pumped through the heart.

Like many space inventions, a big advantage of this new heart monitor was its small size and portability. But it was designed for healthy astronauts and wasn't ready to use on Earth-bound patients with heart ailments. For them, the best way to study the heart was a procedure called a pulmonary artery catheter, in which a doctor would insert a plastic tube into the heart through a small incision near the throat.

A San Diego, California-based company, though, took the NASA technology and developed a system that works for everyone. CardioDynamics International worked with NASA's Johnson Space Center to develop a more sophisticated device called the BioZ® System.

Heart Rate Inc.'s 1-2-3 Heart Rate monitor. Photo courtesy NASA.

It's small, portable and provides 12 measures of the heart, such as how much blood it pumps and how fast blood reaches the extremities.

Besides providing a less-invasive way to measure heart function, the new system can be used in a variety of settings, such as during surgery, for emergency-care patients, or for critically ill patients. Also, it is portable enough to be used in any doctor's office or even in a patient's home. It is simply a small monitor and a few electrodes that attach to a patient's neck and chest. It's a lot cheaper, according to the company: a pulmonary artery catheter costs about $2,000 per test, while a BioZ test costs about $100. The system can be connected to the Internet, which gives medical professionals easy access to a patient's test results. The company is also working on a way to transmit BioZ results from a patient's home to doctors over a telephone line or via e-mail.

The technology helps treat deadly illnesses such as cardiovascular disease and high blood pressure. According to the American Heart Association, cardiovascular disease kills approximately 12 million people worldwide every year, and treatment in the U.S. costs $326.6 billion a year.

Research of space travel and the human heart is continuing. The Houston-based National Space Biomedical Research Institute, formed in 1997 by NASA, is studying the effects of long-term space travel. Researchers know that in microgravity, fluid in the body increases and blood begins to pool in the upper body because gravity isn't pulling it down toward the feet. The extra fluid makes the heart beat faster and even expand in size for a few days. The body eventually rids itself of the extra fluid. After about a week in space, according to NASA's human space flight studies, the heart rate slows, while the cardiac output increases.

Scientists are interested in exactly *when* weightlessness affects fluids in the body and the blood. They think that it may help them understand the other physical changes that happen to humans in space. Knowing what happens out there will help astronauts stay healthy in space, and help them overcome the effects of "cardiovascular de-conditioning" so they can re-adapt to Earth's gravity when they return.

Additional Reading

Kubicek William G., J. N. Karnegis, R. P. Patterson, D. A. Witsoe, and R. H. Mattson. "Development and Evaluation of an Impedance Cardiac Output System." *Aerospace Medicine* 37 (1996): 1208–1212.

"A New Frontier for Cardiac Monitoring." *NASA Spinoff 2001* (2001): 63.

Swenson Jr., Loyd S., James M. Grimwood, and Charles C. Alexander. *This New Ocean: A History of Project Mercury.* NASA Special Publication-4201 in the NASA History Series. NASA: Washington, D.C., 1989.

Heart Pump

A rocket booster and a tool to help ailing hearts would seem to have little in common. Perhaps nobody would have thought of them together if NASA engineer David Saucier hadn't gotten so sick. Saucier worked at the Johnson Space Center in Houston on the massive turbopumps that feed fuel to the Space Shuttle's main engines. In 1983 he had a heart attack, which left his heart badly damaged. Saucier had to wait several months for a heart transplant, which finally happened in 1984, under the care of renowned cardiovascular surgeon Dr. Michael E. DeBakey and his colleague, Dr. George Noon.

While he recovered from surgery, Saucier and DeBakey talked about the challenges of helping patients waiting for heart transplants. DeBakey and Noon had been developing a "ventricular assist device," or VAD, that could help weakened hearts and keep patients alive until a transplant became available. But early versions of the device caused blood clotting and damaged red blood cells. Doctors couldn't figure out a way to move blood through the device smoothly.

Saucier knew NASA engineers had wrestled with similar problems in designing shuttle booster rockets. He brought in scientists from NASA's Advanced Supercomputing Division, and the concepts that helped them move rocket fuel—called "computational fluid dynamics," or CFD—were applied to DeBakey's heart pump. "We could immediately pinpoint what components of the device caused undesirable blood flow patterns," NASA scientist Dochan Kwak said in an interview with *Gridpoints* magazine.

The key part of the pump is a spinning impeller that pushes blood through the device. The curved blades on the impeller spin at 5,000 to 12,000 rotations per minute (rpm), similar to the speed of the low-pressure fuel pump on the Space Shuttle's main engine. Kwak and other NASA scientists recommended changes to the VAD design: Adding an inducer to create smoother blood flow, and changing the angle of the blades to prevent blood from accidentally flowing the wrong direction. Before the changes the DeBakey device lasted just two days in tests on animal subjects. After the changes the device was able to work for more than 100 days.

The improvements didn't come quickly. Saucier had begun working on

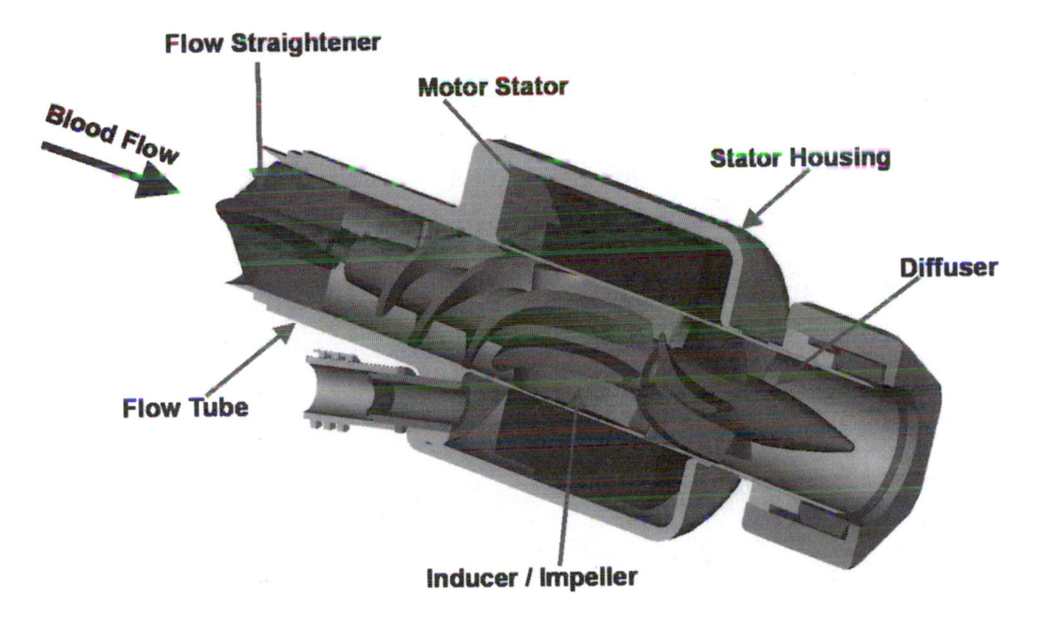

The MicroMed DeBakey VAD. Image courtesy MicroMed Technology, Inc.

the device soon after his transplant, working evenings and weekends along with DeBakey, Noon and other NASA employees and staff from the Baylor College of Medicine. NASA began funding the research in 1992. The device began its development in 1996 when MicroMed Technologies of Houston obtained the license to the technology.

The VAD was first implanted in Europe in 1998. Its first use in the United States came in 2000, when the device was used to help a 31-year-old woman at The Methodist Hospital in Houston. Both DeBakey and Noon took part in the surgery. U.S. clinical trials began in June 2000, and by May 2002 the device had been placed in 151 patients without a single pump failure, according to the company. The 151st patient was a 52-year-old man, who had the device implanted by Dr. Noon in Houston's Methodist Hospital. "This 151st patient underwent four prior surgeries to this one," Dr. Noon said in a company statement. "He had severe heart damage, but with the MicroMed DeBakey VAD, he is now doing just fine. The ease of implantation and the low infection rate is the reason for my decision to use this VAD." In recognition of the success the device has had, in 2001 NASA named the MicroMed DeBakey VAD as its Commercial Invention of the Year.

The device is about one inch by three inches, and weighs less than four ounces. It pumps blood from the left ventricle of the heart through a titanium tube inserted into the heart. The left ventricle has the hardest job, because it has to push blood out to the entire body. After blood passes through the pump it flows to the body through a graft sewn to the aorta. The pump has only one moving part—the inducer, which spins at speeds up to 12,500 revolutions per minute, driven by magnets sealed inside the inducer's blades and two 12-volt batteries carried in a case outside the body.

The device could help thousands of people with end-stage congestive heart failure to live longer. Patients receiving the DeBakey VAD can expect it to last more than five years, according to MicroMed. Approximately 20 million people worldwide suffer from heart failure, according to the American Heart Association, and 5 million of them are in the United States. MicroMed and NASA scientists are still looking at ways to improve the pump by creating controls that would slow it when a patient sleeps, and increase it when they're physically active and need additional blood flow. MicroMed also may develop a version of the device for longer-term use. Plus, Kwak, the NASA scientist, sees a potential for using the NASA-developed computational fluid dynamics analysis to develop other biomedical devices.

Additional Reading

Darwin, Jennifer. "Heart Device Breakthroughs Beating Strong in Houston." *Houston Business Journal*, 28 April 2000, 1A.

"I Have the Heart of a Rocket." *NASAexplores.com* 8 February 2001, http://www.nasaexplores.com/show2_articlesa.php?id=01-005.

MicroMed Technologies Inc. Web site. www.micromedtech.com.

"Miniature Ventricular Assist Device Study Expands." *Medical Industry Today,* 10 January 2001.

"Space Shuttle Engine and CFD Technology Improve Heart Device." G*ridpoints* 1, no. 2 (Spring 2000): 4.

Heart Rate Monitor

The first astronauts in space were well wired under their space suits. Doctors kept track of virtually everything—how fast the astronauts breathed, their heart rate, blood pressure, body temperature, and even how much urine they produced. The data on heart rates was collected through a series of electrodes attached to the astronauts' bodies with a paste-like electrolyte material. The data transmitted back to Earth gave doctors an idea of what stresses the astronauts were going through, which quite often was a significant amount of stress. Take astronaut Gordon Cooper, Jr.'s 34-hour Mercury mission in May 1963 during the early days of the U.S. space program. During his few hours of sleep Cooper's heart rate hovered around 55 beats per minute, according to a NASA post-flight account of the Mercury mission. During the capsule's reentry, though, Cooper's heart was pounding along at 184 beats per minute.

Longer missions, though, required better equipment. The electrolyte paste wasn't suited to long-term use because it eventually dried out, which loosened its contact with the skin and distorted the data. The sensors sometimes produced skin irritations. In the mid-1970s NASA began looking for a new kind of electrode in preparation for long-term missions in space. Researchers at Texas Technical University, working through a NASA grant, developed a new kind of electrocardiographic electrode that had none of the disadvantages of the paste method. The new electrode used a thin dielectric film. The term "dielectric" refers to a material or a medium that can sustain an electric field but not allow current to actually flow through it. That's why the material can be placed against the skin without causing a shock. The dielectric sensors needed no paste and were not affected by heat or cold or perspiration and the signal was not affected by the movement of the electrode against the skin.

The new electrodes were patented by NASA, and then licensed to an inventor,

Richard Charnitski, who in turn used the technology in a line of personal heart monitors and exercise machines sold both for physical fitness and cardiac rehabilitation. Charnitski's company, Heart Rate Inc., of Costa Mesa, California, uses the technology for its 1-2-3 Heart Rate Monitor. The $295 hand-held model works either by touch or with an infrared heartbeat transmitter worn under exercise clothing, and gives users a blinking symbol for each heartbeat and an average heart rate with updates every two seconds. Like its space counterpart the monitor's sensors aren't altered by movement or perspiration.

Additional Reading

"Heart Rate Monitor." *NASA Spinoff 1990* (1990): 126.

Neporent, Liz. "Heart-Rate Hardware." *Shape* 15, no. 7 (March 1996): 54.

Smith, Jeanette. "Long Periods in Space Flight May Take Physiological, Psychological Toll among Crew." *The Journal of the American Medical Association* 263, no. 3 (19 January 1990): 347.

Sonnenburg, Beth. "Monitor Your Progress." *Muscle & Fitness,* August 2001, 116.

Implantable Cardiac Defibrillator

Sometimes a new medical device is so good that regulators shorten the testing period in order to get the product to the marketplace more quickly. That's precisely what happened with a NASA-developed miniature, implantable cardiac defibrillator. The implantable device delivers a shock to the patient's heart when an abnormally fast heartbeat is detected. In this way, the defibrillator works the opposite of a pacemaker, which delivers a steady pulse to help speed up an abnormally slow heartbeat.

The implanted defibrillator works essentially the same way an external defibrillator is used in hospitals. During ventricular fibrillation, the heart stops pumping blood and death or brain damage can occur within minutes.

In the late 1990s, the National Institutes of Health (NIH) conducted a clinical trial with the Automatic Implantable Cardioverter Defibrillator or AICD, manufactured by Cardiac Pacemakers Inc., now Guidant. The study compared the AICD to drugs in terms of effectiveness in slowing down abnormal heart rhythms. The AICD was so successful, the director of the National Heart, Lung and Blood Institute said that more than 1,000 lives, in the U.S. alone, would be saved each year if the device were readily available. The clinical trial was stopped early.

About a half-million Americans die each year from sudden cardiac death (SCD). According to a NASA *Spinoff* article on the defibrillator, about 80 percent of those people die before medical help arrives. Those who survive have up to a 55 percent chance of having another life-threatening heart event within two years. With the AICD, however, that two-year recurrence rate falls to less than two percent.

The AICD uses miniaturized electronics developed by NASA. The development of the internal defibrillator was funded and reviewed by the Applied Physics Laboratory of Johns Hopkins University. The first device was implanted in a human in 1980, four years after an earlier model was successfully implanted in a dog. According to Guidant, more than 65,000 of its devices have been implanted.

While it was known that the devices could help in the instance of a sudden cardiac event, it wasn't until the NIH study years later, when the device was compared to customary drug treatments, that researchers found how effective these devices were in improving overall survival.

The AICD is implanted during a surgical procedure. Current devices, which are smaller than earlier models, are typically implanted under the collarbone—about the same placement of a cardiac pacemaker. The sensor and generator device is connected to the heart by lead wires. These leads provide data on heart rhythm to the sensor and the sensor responds with an electrical shock if the heart rate is abnormal. The shock delivered by a defibrillator has been compared to a kick in the chest.

See also:

Pacemaker Improvements

Additional Reading

"Heart Ticks Right with NASA Technology." *Aerospace Technology Innovations* 5, no. 4 (July/August 1997).

Guidant Web site. www.guidant.com /products/aicd.shtml.

"Implantable Heart Aid." *NASA Spinoffs: 30 Year Commemorative Edition* (1992): 33.

Infrared Detector Improvements

Most objects cannot be seen by the human eye at night or in dark spaces. Even in the daytime, we can only see a small slice of the electromagnetic spectrum. We see visible light—red, orange, yellow, green, blue, violet. On either side of visible light on the spectrum are ultraviolet light (shorter wavelengths than visible light) and infrared (longer wavelengths).

At night, only those objects that glow because they are warm enough, or those that reflect visible light are visible. Yet many of these hidden objects give off an infrared glow. Researchers at the Jet Propulsion Lab in Pasadena, California, working with engineers at Lucent Bell Labs and others, developed Quantum Well Infrared Photodetectors (QWIPs). The ultrasensitive digital infrared sensor was originally developed in the early 1990s to search for distant galaxies and as part of the "Star Wars" defense program to spot missile launches. Since then, the technology has resulted in one of the most sensitive handheld infrared cameras—with multiple applications both in space research and commercial goods.

The infrared spectrum is divided into measurements called microns. For example, an 8.5-micron infrared detector is typically used in security and surveillance, navigation and flight control, and early-warning systems. These cameras also have uses in medical imaging applications because they can map the surface of skin. Telescopes equipped with 8- to 12-micron detectors are able to look at distant stars and galaxies.

Infrared imaging systems that can "see" the absorption of molecules of ozone, water, carbon monoxide, carbon dioxide, and nitrous oxide can monitor changes in global weather profiles, pollution, deforestation and other atmospheric changes on Earth.

The QWIP technology, which has been used in space to help NASA observe the Earth more clearly, is less expensive than existing infrared imaging technology. It can also reach longer infrared wavelengths than previous technologies could.

JPL licensed the technology to a new company, QWIP Technologies, to produce focal plan cameras. It has been used for a variety of purposes.

- As a fire-observing device. In October 1996, a Los Angeles TV news crew attached a QWIP infrared camera to their

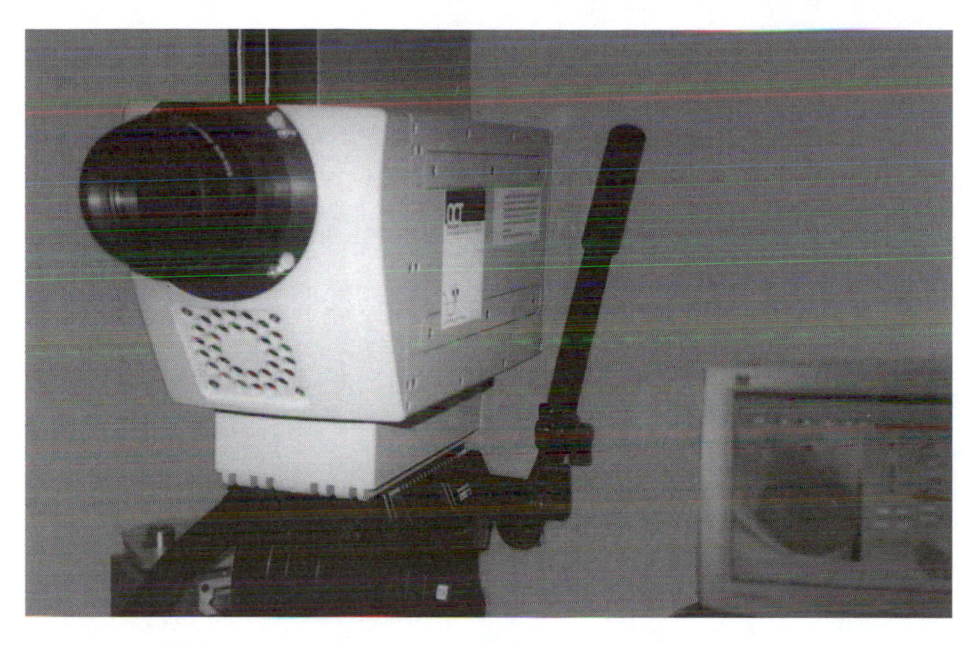

OmniCorder Technology Inc.'s QWIP sensor. Photo courtesy NASA.

news helicopter. Not only could the camera "see" through the smoke, it also helped firefighters identify hot spots—areas that were likely to go up in flames again even though it looked like the fire was out. The camera was effective during the day light and at night.

- For monitoring volcanoes, mineral formations, weather and atmospheric conditions. In 1997, the QWIP camera was taken to the Kailua volcano in Hawaii. It was able to detect features at much higher temperatures. For example, the QWIP camera images clearly showed an underground hot lava tube running up the mountain—something that was not visible to the eye or even earlier infrared devices.

It also has been used for medical applications. A biomedical company, OmniCorder Technologies, Inc., has licensed the technology for medical imaging purposes—most notably breast cancer detection. The company uses as its slogan on its Web site, "Using Aerospace and Defense Technology to Solve Healthcare Problems." In a January 2000 press release from JPL, OmniCorder received clearance from the U.S. Food and Drug Administration (FDA) in December 1999 to market its BioScan System, which uses the QWIP detectors. "Studies have determined that cancer cells exude nitric oxide. This causes changes in blood flow in tissue surrounding cancer that can be detected by the sensor. The BioScan System is sensitive to temperature changes of less than .015 degree Celsius (.027 degree Fahrenheit) and has a speed of more than 200 frames per second. It causes no discomfort to the patient and uses no ionizing radiation," the press release states.

Not only does the system provide a non-invasive way to diagnose cancers, it is also being used to monitor cancer treatments. At Boston's Dana-Farber Cancer Institute, researchers and physicians began using the system to determine if there are any changes in blood supply to cancerous lesions following treatment with anti-cancer products being tested.

In other medical applications, the camera was also tested by researchers at the Texas Heart Institute in an experiment on a heart surgery on a rabbit heart. QWIP detectors were able to see arterial plaque built inside the heart.

Additional Reading

"Breast Cancer Screening Aid Cleared for Diagnostic Use." NASA press release, Jet Propulsion Lab, Pasadena, Ca. 28 January 2000.

Flinn, Edward D. "From Star Wars to a War against Cancer." *Aerospace America* 38, no. 17 (July 2000): 22.

"The Great Government Giveaway: Grants for Small Businesses That Use Military Technology." *Business Week*, 12 June 2000, F42.

"Leading Cancer Institute Tests Novel Monitoring Technique." NASA press release, Jet Propulsion Lab, Pasadena, Ca. 30 March 2000.

Omnicorder Technologies Web site. www.omnicorder.com.

QWIP Web site. www.qwip2000.jpl.nasa.gov.

Infrared Thermometer

What do distant stars and a child's fever have in common? The way they're measured. NASA's Jet Propulsion Lab (JPL) developed a system more than 30 years ago to measure the amount of infrared energy emitted by stars and planets. NASA used this technology to study stars in our Milky Way galaxy and beyond.

In 1991, Diatek Corporation, a Welch Allyn company brought the same technology into everyday use in hospitals and nursing homes. By then, the country was facing a national nursing shortage. The company began looking for quicker ways of getting an accurate temperature reading on patients. When a nurse can take a temperature in a few seconds compared to 30 or more, that adds up to many saved minutes over the course of a busy day. Diatek researchers, with help from JPL scientists, developed the Diatek Model 7000 aural thermometer. The handheld device weighs eight ounces and can measure temperature in less than two seconds by using the same technology to measure the infrared energy of the eardrum. (Since Diatek was acquired by Welch Allyn, the thermometer's name was changed to SureTemp. Welch Allyn's newest line of infrared aural thermometers are in the InstaTemp line and in 2001, the company agreed to market the Braun ThermoScan® Pro 3000 Ear Thermometer to the medical professional market.)

As a result, more patients can be seen, or more time and attention can be given to the patient's other needs. The frequent temperature readings taken on ill or newborn patients are much less of an inconvenience. It is well suited for hospitals, particularly newborn and children's units, nursing homes, and physicians' hospitals.

Using the device is simple, as explained in NASA's *Spinoff* magazine. The nurse "inserts the plastic-covered probe into the opening of the patient's ear canal and presses a button to activate the sensor. The probe detects infrared radiation emitted from the eardrum and a microprocessor converts it to the corresponding body temperature, which is displayed on a liquid crystal screen." The disposable plastic covering is replaced with each patient, virtually eliminating any risk of cross contamination.

While health care practitioners have appreciated the simplicity of this and other infrared aural devices, some studies have shown that they are not always accurate. As a result, patients can be over treated or under treated. An incorrect temperature reading can be especially problematic with infants, where treatment often begins at specific fever levels. One 1998 study showed that infrared aural thermometers often gave different readings for each ear, even when the measurements were taken within seconds of each other, and the infrared reading varied significantly from the oral mercury thermometer reading in about one-third of the cases.

Even so, infrared aural thermometers have expanded to the home market. Some of the differences in readings can be due to positioning the device and obtaining a good seal with the opening to the ear. Still, parents often find them convenient screening tools for determining their child's temperature range. On online parent and medical sites, parents are typically urged to double check an aural thermometer's reading with an oral or rectal thermometer reading if they are not wholly confident in the aural reading.

Additional Reading

"Diatek Ear Thermometer to Hospitals, Alternate Sites." *Health Industry Today* 54, no. 6 (June 1991): 17.

"Infrared Thermometer." *NASA Spinoff 1991* (1991): 80.

Modell, J.G., C.R. Katholi, S.M. Kumaramangalam, E.C. Hudson, and D. Graham. "Unreliability of the Infrared Tympanic Thermometer in Clinical Practice: A Comparative Study with Oral Mercury and Oral Electronic Thermometers." *South Medical Journal* 7, no. 91 (1998): 649–54.

Insulin Pumps

Type 1 diabetes is a disease in which the body can no longer produce insulin, a hormone required to convert sugars and starches from the food we eat into the energy we need. People with the more common Type 2 diabetes are still able to produce insulin, but either the body produces too little or has become resistant to it. Type 2 diabetics, who account for more than 90 percent of diabetes cases, can often control their blood sugar levels through weight loss, exercise and better nutrition. Some may need oral medication or insulin injections.

For the 500,000 to one million people in the U.S. with Type 1 diabetes, however, insulin injections are essential. Without them, blood sugar levels would spike to fatally high levels. Most Type 1 diabetics require multiple daily injections. Typically, the person calculates the amount of carbohydrates to be eaten and injects a certain number of units of insulin to counteract the resulting rise in glucose levels. Injecting insulin before every meal or snack is the most targeted approach to diabetes treatment and allows the greatest flexibility, but some diabetics find it cumbersome. Another approach is to inject a specific amount of insulin in the morning and then eat only those foods that will be covered by that amount of insulin. The second approach is more regimented and doesn't allow for spontaneous snacking.

NASA technology has given rise to two major developments in insulin delivery—an external infusion pump and an implantable pump. Both allow a precise delivery of insulin as directed by the diabetic—without individual injections.

MiniMed Technologies developed an insulin infusion pump in 1985. The 504 model incorporated many improvements over the company's earlier rudimentary pumps. The pump itself is in a small case—about the size of a small beeper—and weighs less than four ounces. Besides the pump and the programmable microprocessor, the device contains a syringe containing several days' worth of insulin.

The syringe is connected to an infusion set—thin plastic tubing with a needle at the end. The diabetic inserts the needle, typically in the abdomen, and it stays there until the syringe is refilled a few days later. The user directs the microprocessor to drip insulin at a low dose around the clock and can also program it to give a larger dose before meals. As a result, pump users only need to inject themselves once every two or three days rather than multiple times each day.

MiniMed spun off from an earlier company called PaceSetter Systems. In 1980, PaceSetter Systems, a cardiac pacemaker manufacturer, teamed up with NASA and the Applied Physics Laboratory at Johns Hopkins University to work on an insulin pump. Five years later, PaceSetter spun off the insulin pump business as MiniMed Technologies Limited.

During this time, NASA and MiniMed had also been working on the Programmable Implantable Medication System (PIMS). The result was an implant, about the size and shape of a hockey puck, which contains a supply of insulin that lasts about three months. The microprocessor, pumping mechanism and refillable insulin reservoir are implanted in the abdomen. A catheter leads from the pump directly to the diabetic's intestines to deliver the insulin.

The half-pound pump is made of titanium. As Johns Hopkins physician Christopher Saudek, M.D., explained in a 1989 article in *Diabetes Forecast,* the magazine of the American Diabetes Association, "The body cannot reject the pump the way it does a transplant, because it's made of titanium. The normal response is to wall it off with a thick fibrous tissue that has no nerves in it—a kind of pocket."

Years of work by NASA to miniaturize components for use on satellites made this small, implantable pump a reality. The pumping technology itself was another NASA contribution. It was based on a design used in the biological lab of the Mars Viking space probe. By delivering the insulin in short bursts, the device saves battery power. The pump uses a 3.6-volt battery that should last more than five years. NASA's work on communication was the key to how the device can be programmed. NASA uses bi-directional telemetry to communicate between spacecraft and Earth. This telemetry has been used in several medical devices (*see* "Pacemaker Improvements"). A doctor or the diabetic can hold a small radio transmitter over the implant and change the infusion rate or direct the pump to deliver a supplemental dose of insulin. Doctors can gather information from the device's memory to help better guide a patients' care.

In November 1986, the first PIMS was implanted in a patient at the Johns Hopkins Hospital. By 1990, the implantable pump was being tested widely and it received market approval for use in Europe in 1995.

In 1988, MiniMed was inducted into the Space Technology Hall of Fame for using NASA technology in this application.

In June 2002, MiniMed, now Medtronic MiniMed, reported that three studies showed that for many patients, "pump therapy is the most effective way to maintain tight glucose control, and can result in substantial clinical benefits, such as fewer diabetes-related complications."

Additional Reading

Budiansky, Stephen, Joseph Carey, and Stanley N. Wellborn. "Out of the Lab, into the Real World." *U.S. News & World Report,* 29 December 1985, 86.

"Implantable and External Pumps." *NASA Spinoffs: 30 Year Commemorative Edition* (1992): 34.

"Insulin Pumps Can Reduce Health Risks Associated with Diabetes Management." Medtronic MiniMed press release, 17 June 2002.

Mazur, Marcia. "A Prime Pump." *Diabetes Forecast* 42, no. 3 (March 1989): 24.

Medtronic Minimed Web site. www.minimed.com.

"NASA-Style Pump Is Implanted in Diabetic." *New York Times,* 19 November 1986, A20.

Joysticks for Computers

Playing computer games today is a whole lot more fun—in large part, thanks to NASA. That's because NASA needed to have realistic simulation training for astronauts. The key to that was the development of the Rotational Hand Controller (RHC), essentially a joystick. The RHC was developed by ThrustMaster for Lockheed Martin and the Johnson Space Center. It's used by astronauts to practice maneuvering during flight as well as landing techniques. It is also used onboard the shuttle so astronauts can simulate shuttle landings while still in orbit.

ThrustMaster, Inc., of Hillsboro, Oregon, has taken many of the elements of the NASA RHC and brought it into a commercially available line of computer game controllers—joysticks for flight simulators and auto racing controllers for driving and racing simulators. The flight simulator, released in 1997, a year after the joystick, is a 3D, programmable controller.

At about the same time that ThrustMaster was developing the RHC, another company was also working to add a sense of realism to computer joysticks. Louis Rosenberg, a Stanford University engineering graduate student, wanted to bridge the gap between the visual senses and the sense of touch. His experimental lab turned out to be the NASA Ames flight simulator with its force feedback joystick. NASA has long been studying haptics—the science of feel—to enhance the remote technology needed in space missions. The $100,000 dishwasher-sized machine provided a sense of touch to the user—plus it was a great deal of fun to manipulate. So Rosenberg launched Immersion Corporation in 1993 to bring affordable force feedback technology to computer games. Now Immersion makes a feel chip that allows game controllers such as steering wheels or joysticks to vibrate, feel resistance and spring, and otherwise thump, bump and jar.

The company has moved beyond the gaming industry, however. With its TouchSense technology available to software makers, users with a touch-enabled mouse can "feel" what's on their screen. There may be a bump or a jump when the cursor hits an icon; there might be the feeling of hitting a wall when a control is moved all the way to its end. In 2001, Immersion Corporation announced an agreement with educational software maker The Learning Company to include TouchSense in its children's software. In Immersion's press release announcing the development, the company said the sense of touch could improve a child's mouse navigation, as well as add a new level of fun to educational games.

Immersion Corporation has also spun off into the medical field—another area where simulated training benefits from a sense of touch. The company's medical subsidiary, Immersion Medical, makes computer-based medical simulation systems. In May 2002, the company received a grant from the National Institutes of Health to develop a hysteroscopy surgical simulator. During a hysteroscopy procedure, the surgeon can look into the cavity

Joystick. Photo courtesy NASA.

of the uterus with a thin instrument and can correct uterine abnormalities, remove tumors and even remove the entire uterine lining. However, as the press release announcing the grant states, surgeons don't have a true feel for what the procedure will be until they perform their first one. "Each year more than 229,000 hysteroscopies are performed on women in the United States. Currently, hysteroscopy is learned on patients or inanimate objects such as bell peppers or sheep bladders. Using inanimate objects prevents trainees from feeling realistic sensations when they manipulate the tools, and does not allow objective feedback on their performance."

Another of Immersion's recent developments includes automobile controllers. The new BMW 7-series autos have the new iDrive technology, putting 700 comfort and convenience controls into one hand-sized, knob-like controller positioned between the front seats. The idea is to provide drivers with one intuitive control, rather than a dashboard full of knobs and buttons.

The need to allow astronauts to simulate maneuvers became increasingly acute after a June 1997 accident in space. While no one was injured, an unmanned cargo craft containing garbage collided with the Russian Mir space station during a docking test. As a result, the space station lost some of its air pressure. Since then, NASA Ames Smart Solutions Group researchers developed a 3D interactive Space Shuttle/space station docking simulation.

Matched with a computerized joystick enabled with force feedback, astronauts can practice these risky maneuvers before attempting them in space—where repairing damage is difficult and dangerous.

Finally, one other major application of the joystick technology has nothing to do with computer gaming, but has had an enormous impact for people with disabilities. The same joystick controller developed by NASA in the 1970s as a control system for the Lunar Rover has been used in automobiles and vans to allow drivers to control a vehicle with one hand. Called a Unistick, the technology was developed by Johnson Engineering of Boulder, Colorado, in 1986.

Additional Reading

"Force Feedback Joystick." *NASA Spinoff 1997* (1997): 77.

Gallagher, Leigh. "Feel Me." *Forbes,* 20 March 2000, 278.

Hines, Michael. "Simulator Offers Virtual Viewpoint." *Knight-Ridder/Tribune News Service*, 9 April 2002.

Immersion Corporation Web site. www.immersion.com.

"Joy of a Joystick." *NASA Spinoff 1998* (1998): 72.

"NASA Simulates Space Shuttle Docking." *Computer-Aided Engineering* 18, no. 10 (October 1999): 10.

"Once More, with Feeling." *Stanford University School of Engineering, Annual Report, 1997–98.* http://soe.stanford.edu/AR97-98/rosenberg.html. Accessed 8 June 2003.

ThrustMaster Inc. Web site: www.thrustmaster.com.

"Vehicle Controller—Unistick." *NASA Spinoff 1985* (1985): 75.

Knee Brace Improvements

The human knee is a complicated device in its own right. Comprised of four bones, it helps control how fast a person can move, how much energy can be applied from muscles when lifting objects as well as the direction that energy travels. When it's not working properly, the impact on a person's mobility is enormous.

Oftentimes the only answer for a knee weakened by injury or illness has been wearing a brace, but that was at best a partial solution. For years, most braces available were those that locked the knee joint into a rigid, straight position. While that made the leg able to support weight, it didn't give the patient any kind of normal gait. Plus, it didn't allow any kind of strengthening or rehabilitation of the knee joint for persons recovering from an injury.

The technology for a new kind of knee brace came in the mid-1990s from a team of engineers at NASA's Marshall Space Flight Center in Huntsville, Alabama. They created a flexible knee brace that allows the wearer to move the knee joint, but still provides the strength to support the user's weight. Here's how it works: The top of the brace attaches to the patient's thigh, and the lower part is secured to the heel. When the patient puts weight on the heel, as they would during a normal walking stride, the brace locks in position and takes strain off of the injured knee. When weight is removed from the heel, the brace becomes flexible again. That flexibility also makes it easier for patients to rehabilitate an injured knee and speed their recovery. The brace is also compact enough to be worn underneath clothing.

According to principal inventor Neill Myers, the team drew on several space-based technologies. One was the design of a zero-gravity clamping device that Space Shuttle astronauts could use with one hand to transfer new batteries to the Hubble

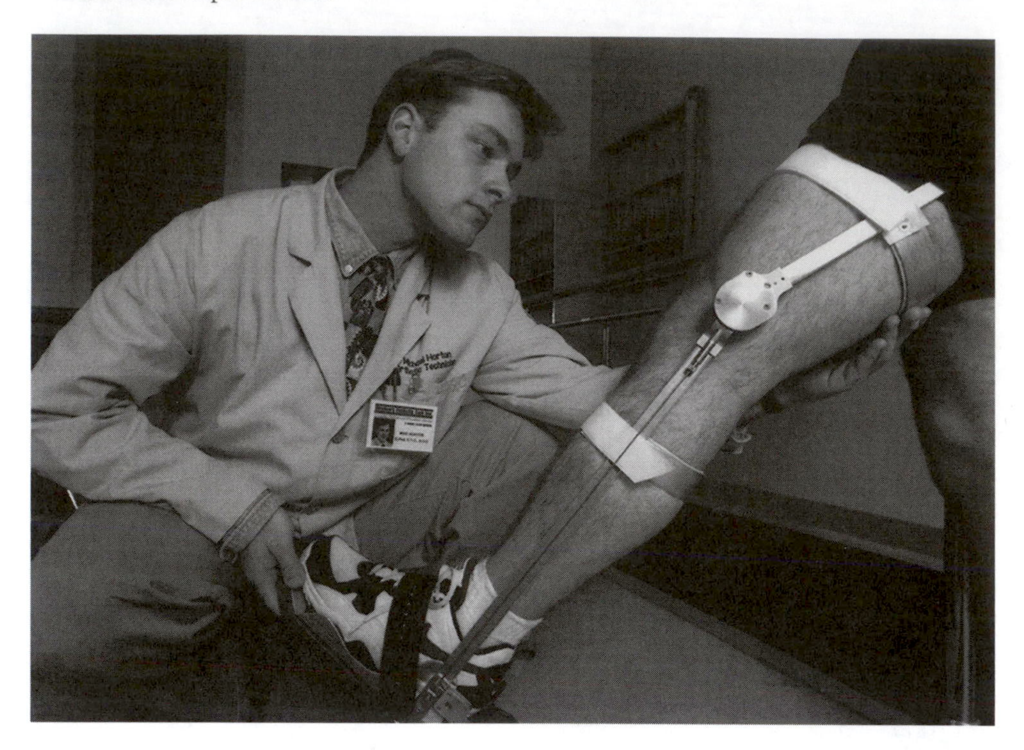

Horton Orthotic Lab's Stance Control Orthotic Knee Brace. Photo courtesy NASA.

space telescope during a servicing mission. A second was the design for a space vehicle rendezvous docking mechanism. When two spacecraft have been maneuvered into close proximity, they must be securely joined together. This particular docking mechanism, installed in one of the vehicles, captures a mating bar on the other vehicle, pulls it in, and locks it down, firmly securing the vehicles together. Another propulsion-related project that provided background expertise for the knee brace engineering team was the design of actuators for rock engine thrust vector control systems, which gimbal or pivot, a rocket engine to provide steering capability for the rocket booster or spacecraft.

The engineers who designed the brace didn't start the process on their own. Private laboratories that specialize in designing orthotic devices asked the Marshall Center for help, and it assembled a team of engineers trained in propulsion technology and motion control. The Marshall Center designed the Mercury-Redstone booster that launched the United States' first astronaut, Alan Shepard, into space in 1961. It has also produced the Saturn rocket booster systems that powered the Apollo moon missions, as well as the boosters that have carried the Space Shuttles into space since 1981. Besides Myers, others on the design team were co-inventors Michael Shadoan, John Forbes, Kevin Baker and Darron Rice.

The team also worked with Horton Orthotic Lab of Little Rock, Arkansas to get the brace ready for commercialization. The Horton Company licensed the product from NASA in 1997. It continued to refine the brace to make it ready for use, and did clinical trials in 2001 with six patients. The trials were successful; and in 2002 the company began making it available to patients. Horton calls it the "Stance Control Orthotic Knee Joint." The demand for orthotic devices is sizable. About five million people visit orthopedic surgeons each year because of knee problems, according to the American Academy of Orthopaedic Surgeons. About three million of those are due to injuries; the remaining two million are because of other disorders, such as arthritis. NASA engineers originally designed the brace for knee injury and stroke patients, but Myers says it can also be used to help persons with spinal cord injuries and birth defects such as spina bifida.

Additional Reading

Brown, Jean Park. "Orthopaedic Care of Children with Spina Bifida: You've Come a Long Way, Baby!" *Orthopaedic Nursing* 20, no. 4 (July 2001): 51.

Global Resource for Orthotics and Prosthetics Web site. www.oandp.com/news/jmcorner/2001-10.asp John Michael's Corner.

Myers, Neill and Gary Horton. "Space-age Bracing." *BioMechanics,* July 1998, 12–16.

"Quicker Rehabilitation for New Knee Brace Wearers." *Innovation* 5, no. 1 (January/February 1997).

Landmine Removal

The two 149-foot-tall solid rocket boosters (SRBs) attached to the Space Shuttle at launch are the biggest solid-propellant motors ever flown. Ignited just after the shuttle's three main engines, they provide 71 percent of the shuttle's power during lift-off and during the first stage of its ascent. They are massive, powerful engines. Each weighs about 1.3 million pounds at launch—and nearly all of that weight is the propellant itself, a mixture of ammonium perchlorate, aluminum, and a polymer to hold it all together. The SRBs together push the shuttle 24 nautical miles into space, about 150,000 feet, before separating from the shuttle and falling back to Earth by parachute, where they are recovered and reused.

When the propellant for the SRBs is mixed, a little extra is prepared to make sure enough is available for launch. That surplus fuel, though, soon turns to a solid, and can't be used for the next mission. But it doesn't go to waste. The company that makes the propellant, Thiokol Propulsion of Brigham City, Utah, uses it to produce a device than can destroy landmines where they are buried.

Thiokol uses the propellant in a small flare, which is easy to use and safe to handle. The flare is mounted on a three-legged stand and is only about 13 centimeters (5 inches) long. When placed next to an uncovered landmine and ignited from a safe distance with a battery-triggered electric match, the heat of the rocket fuel burns a hole through the casing and ignites its contents. The flare burns for about 70 seconds at a temperature of 3,500 degrees Fahrenheit. Sometimes the mine explodes, but even then the explosion is less than its full force and inflicts less damage, Thiokol employee Charles Zisette told NASA. It's a lot safer than deactivating the mine by hand, or by simply detonating it. Also, because the flares are simple to use and easy to transport they can be carried to remote locations where larger demining equipment can't go. The fuel is free, so it costs the demining campaigns around the world nothing, and it reduces the amount of fuel that has to be disposed of, avoiding those environmental impacts as well. The flares are sold for about $10 each, according to the U.S. Department of Defense's Humanitarian Demining Developmental Technologies 2000–2001 catalog.

Thiokol Propulsion's anti-landmine flare. Photo courtesy NASA.

The Thiokol device has been tested against eight different types of mines. According to the de-mining technologies catalog, the flares were able to penetrate 2 millimeters of steel and 10 millimeters of wood. Tests done in December 1999 showed the flares generally set off a mine's detonator within 25 to 65 seconds. The explosion that the flare produced, though, was so weak that the flare stand wasn't knocked over and no crater was created. The flares didn't fare as well against mines with casings thicker than 4 millimeters, however.

In 2000, the United States Department of Defense provided Thiokol flares to the Lebanese Armed Forces to help it clear landmines and unexploded ordnance. A series of wars in Lebanon in the 20th century have left the country with an estimated 280,000 mines and unexploded ordnance, nearly half in south Lebanon near the border with Israel. Many of the minefields are marked, but some are not. As of July 2000, landmines and unexploded ordnance had killed 1,168 people and wounded 1,546 more, according to the U.S. Department of State. Many of those victims were farming, which is the main income source for villagers in south Lebanon.

The flare was conceived in collaboration with DE Technologies, Inc. of King of Prussia, Pennsylvania, and the Marshall Space Flight Center in Huntsville, Alabama. Marshall is NASA's lead center for developing space transportation and propulsion systems. Worldwide, there are an estimated 80 million or more active landmines, scattered in at least 70 countries, according to NASA. Landmines kill or maim 26,000 people a year.

The Thiokol flare is not the only demining device developed from space research. A North Carolina company, hired by NASA to develop an airborne electromagnetic sensor, also developed a technology it uses to make a landmine-detecting device. The Geophex company's GEM-3, which looks like a big metal detector, is a broadband digital electromagnetic sensor that can detect landmines. It weighs about ten pounds and uses two electromagnetic coils to make a magnetic field. Objects in that field have unique electromagnetic responses, and that makes it possible to identify the kind of landmine or unexploded ordnance buried in the ground. Knowing what kind of explosive is buried at a site without having to dig it up makes it safer and easier to decide how to dispose of it.

Additional Reading

International Campaign to Ban Landmines Web site. www.icbl.org.

"NASA Fuels Landmine Removal Efforts with Thiokol." *Journal of Aerospace and Defense Industry News,* 5 November 1999, 41.

NASA Human Space Flight Space Shuttle reference Web site. http://spaceflight.nasa.gov/shuttle/reference/index.html.

"To Walk the Earth in Safety: The United States Commitment to Humanitarian Demining." Report by the U.S. Department of State. November 2001.

U.S. Department of Defense's Humanitarian Demining Developmental Technologies 2000–2001 Catalog Web site. http://humanitarian-demining.com/catalog/contents/fcover.htm.

Laptop Computers

The first laptop computers were not built for space. Made in the late 1970s and early 1980s, they were simple and heavy models without much memory. The first users were a few business executives and government civilians, along with the military, which wanted portable computers that were rugged enough to take into the field and were simple to operate. These new tools also caught the eye of planners of the Space Shuttle missions, who wanted a portable computer—separate from the shuttle's main computer system—to provide astronauts with a small-screen image identical to the large-screen projection of the shuttle's course and position as seen at the Mission Control Center in Houston.

The same standards that applied to all the equipment carried into space were applied to choosing a portable computer. It had to be lightweight and yet strong enough to withstand the force of a launch. It also needed to be able to display graphics and have a high processing speed and sufficient memory. From the few models available on the market NASA chose the Compass, a 10-pound, black notebook-sized computer made by Grid Systems Corp. of Fremont, California.

Before it could go into space, though, the Compass laptop underwent improvements in design and software—changes that helped create a new generation of consumer-oriented laptops. NASA engineers added a fan to cool the components. The original model was cooled by simply letting heat escape through the circulation of air, but that wouldn't work in a weightless environment. NASA's programmers also added a larger electroluminescent screen and rewrote the source code for the computer. The company later credited NASA's input on the computer "because it helped fine-tune the Grid operating system and common code documentation," according to the *NASA Spinoff 1985* publication. The Grid—called the Shuttle Portable Onboard Computer—was first used on a nine-day Space Shuttle mission in November 1983.

Like many emerging technologies, these first laptops were expensive. The Grid Compass cost $8,000. *Government*

Computing News magazine in 1989 called them "a computing status symbol of the early 1980s. You had to be pretty high up to rate one." The Compass used the same clamshell-style design seen in modern laptops, with the screen folding down over the keyboard. Its operating system was built around the Intel 8086 chip, and didn't use Microsoft DOS. It also used bubble memory, an early kind of computer memory in which data is stored in circular areas on a thin film of magnetic silicate. The information isn't lost when the computer is turned off, and in the early days of portable computers, it was believed to be the best choice for memory. Eventually, though, bubble memory was replaced by faster and less expensive memory technologies. The Compass model had 384K of bubble storage.

Additional Reading

Alexander, Michael. "SPOC Flies on the Shuttle." *Computerworld* 22, no. 46 (14 November 1988): 61.

"Compass Points the Way for Business Uses." *PC Week* 2, no. 4 (29 January 1985): 109.

Margerum, Barry. "GRID Locks Up Washington." *PC Magazine* 4, no. 15 (23 July 1985): 62.

"Portable Computer." *NASA Spinoff 1985* (1985): 104.

Robb, David W. and Cynthia Morgan. "Looking at Micro Classics." *Government Computer News* 8, no. 25 (11 December 1989): 1.

Laser Angioplasty

The idea of using lasers in space conjures up images of futuristic weapons. While that may someday be true, today lasers in space are far less threatening. NASA has been using lasers for its study of the Earth's atmosphere. Scientists at the Jet Propulsion Laboratory developed a technology that could send pulses of laser-generated light into the atmosphere, where it could measure things like ice crystals, aerosol particles and trace gases such as ozone. Learning the altitude and location of these kinds of atmospheric components allowed scientists to develop more advanced weather forecasting models and to monitor the environment.

The laser worked the same way that sonar uses sound waves to detect objects. It sends out laser pulses and measures the light that bounces back. Known as LIDAR, for light detection and ranging, the technology also makes it possible to measure wind speeds or track plumes of pollution.

The pulsating laser that NASA developed had another characteristic: The powerful beam it produced was cool, not hot. The laser was capable of producing ultraviolet light with a temperature of just 65 degrees Centrigrade, a temperature that human tissue could withstand. In the late 1980s, medical researchers began a series of clinical tests using the pulsating laser technology to attack clogged arteries using a new device called an excimer laser. It is a gas laser, powered by xenon chloride gas. Excimer is an acronym for excited dimer—and refers to the short-lived molecules that emit the ultraviolet radiation when energized into an excited state.

Inserted into the clogged artery through a flexible catheter of fiber optic bundles, the laser's beam vaporized fatty deposits that built up in arterial walls without damaging healthy tissue. In the clinical trials it had fewer complications than balloon angioplasty, a common nonsurgical treatment in which a balloon-like device was inserted to the blockage and inflated to allow blood flow. Plus, it is far less invasive than the option of bypass surgery, which was the principal method for treating arterial blockages that couldn't be cleared through balloon angioplasty.

The initial tests of the excimer laser system, done in the late 1980s, involved nearly 5,500 patients in two major studies, and the success rate was between 85 percent and 90 percent. In January 1992, the system received Food and Drug Administration approval for treatment of coronary disease.

The first commercially produced excimer laser was made by Advanced Intraventional Systems of Irvine, California. Its Dymer™ 200+ Excimer Laser system used the technology NASA developed to produce a laser beam that can produce pulses as low as 200 billionths of a second, which is how it maintains such a low temperature.

Albert Einstein was the first to come up with the idea for a laser, in 1905. Laser is an acronym that stands for light amplification by stimulated emission of radiation. While studies have shown that excimer lasers work to vaporize plaque buildup in arteries, they aren't exactly sure how it works. Researchers suspect the process probably depends on breaking down the unwanted molecules through photochemical reactions triggered by the laser.

A Colorado company, Spectranetics, bought Advanced Intraventional Systems, the original maker of the laser angioplasty system, in 1994. Its current system, the CVX-300, is the only device approved by the FDA for treating multiple cardiovascular diseases. It can clear blockages in the heart or legs, and it can be used to eliminate the scar tissue that can build up around pacemakers and the thin strands of wire used with implantable cardioverter defibrillators (ICDs). Patients who undergo laser angioplasty are typically given a mild sedative but are awake for the procedure, according to Spectranetics literature. The laser catheter is inserted through a site near the groin or on an arm, and the patient and medical staff wear special glasses to protect against the laser's UV light.

Excimer lasers are also used to correct vision problems. Known as LASIK surgery, the process involves using the laser's ability to vaporize tissue on the stroma, which is the mid-section of the cornea. By changing the shape of the cornea, surgeons can improve the cornea's ability to focus and correct problems such as myopia, or nearsightedness, or hyperopia, also called farsightedness.

Additional Reading

Ebersole, Douglas G., MD. "Clinical Applications for the Excimer Laser." *Cardiovascular Reviews & Reports: Worldwide Advances in Disorders of Blood Pressure, Heart, and Circulation* 20, no. 6 (June 1999).

Jet Propulsion Laboratory Remote Sensing Group Web site. www.jpl.nasa.gov/lidar/welcome.htm.

"Laser Angioplasty." *NASA Spinoff 1990* (1990): 74.

Spectranetics Web site. www.spectranetics.com/cvx/cvx300.html.

U.S. Food and Drug Administration Center for Devices and Radiological Health Web site. www.fda.gov/cdrh/lasik.

Memory Foam

Spend a half hour in an uncomfortable chair and your body will start complaining. Imagine if that chair were your seat as you blasted off into space. One of the challenges of space researchers has been to design comfortable seating for astronauts as their bodies are pressed back by incredible gravitational force during takeoff and descent. What they were looking for was a material that would mold to the body of the astronaut, creating a customized fit.

What researchers at NASA's Ames Research Center developed in the 1970s was an open-celled, temperature-sensitive viscolelastic material that was able to distribute body weight evenly. The problem was that it didn't work very well in space because the foam became stiff in the very cold temperatures. However, the material called memory foam has become one of the most successful NASA spin-off ventures ever created.

When NASA released the technology in the early 1980s, Fagerdala World Foams, a Swedish foam manufacturer, began experimenting with the material. It took nearly 10 years of research and development until the company released its Tempur material and incorporated it into a mattress design. The pressure-relieving mattresses were first tested in hospital.

According to a 1998 press release from NASA, the material has been shown to reduce bedsores in patients. The press release stated that "Both the Veterans Affairs Department and the National Institutes of Health have purchased hundreds of Tempur-Pedic products for use in their pain management and ulcer treatment programs. Bedsores, which can be fatal if left untreated, cost the Medicare and Medicaid programs almost $2 billion annually for treatment of wheelchair-bound, nursing home and hospital patients."

In 1991, the company released Tempur-Pedic Swedish Mattress in Sweden and released the same products the following year in North America. On the company's Web site, it explains how this memory foam contributes to comfort. "In warm areas, [the cells] get softer and pliable. In cooler areas, they stay firm. The cells will literally shift position and reorganize to conform to your body contours. The cells shift so we don't have to. Imagine a mattress that's firm where you need it and soft where you want it. It's like having a mattress custom designed to fit your body."

For example, people who sleep on their backs find that the heaviest part of their body—their hips—sinks down further than their shoulders or feet. The Tempur material actually rises to fill the gaps so that the person's neck and spine are aligned.

According to the Tempur-Pedic Web site, research has proven that this system inspires a better night's sleep. The Web site mentions that research has been conducted at the Institution for Clinical & Physiological Research at Lillhagen Hospital in Gothenburg, Sweden and the University of Maine. "Clinical studies indicate that the Tempur-Pedic mattress cuts the average nighttime tossing and turning by more than 70%," according to the site.

NASA still uses the material as cushioning in Space Shuttles. In addition, the memory foam material has been used in pillows, home and office furniture, bicycle seats, mattresses covers and seat cushions, operating table pads, football helmets and earplugs—in other words, any surface on which a person needs to sit or sleep for long stretches of time.

Tempur-Pedic became one of the first companies to receive approval to use the "Certified Technology" Seal on its products and advertising. Through the Certified Space Technology initiative, founded by *Apollo 13* commander Jim Lovell, products or services certified as having originated in U.S. space efforts or used in flight may use this seal.

Additional Reading

Diamond, Nina L. "From Outer Space to You: Turning NASA Research into a Comfy Chair." *Omni,* March 1994, 24.

"Famous Foam Has a Future." *Aerospace Technology Innovation* 6, no. 3 (May/June 1998). Information also available on NASA Commercial Technology Web site. http://nctn

.hq.nasa.gov/innovation/Innovation63/foam .htm.

Space Connection Web site. www.space connection.org.

"Space Race to Marketplace: Spinoffs from Space Technology Land on Earth." *The Futurist* 27, no. 5 (September–October 1993): 53.

Starkey, Danielle. "NASA's Failed 'Memory Foam' Finds New Mission." *San Francisco Business Times*, 21 September 2001, 25.

Tempur-Pedic Web site. www.tempurpedic .com.

"Test Pilots to Testbeds—NASA Cushions Liftoff and Eases Bedsores." NASA press release, 6 May 1998.

Memory Metals

Eyeglass frames that can fix themselves when bent, water faucets that shut off if the water gets too hot, and bendable surgical tools: these are just a few of the uses for a remarkable kind of metal alloy with what scientists call a shape memory effect, or SME. What that means is the metal can remember a given shape—and return to that shape when exposed to certain stimuli, such as heat, electricity or pressure.

Shape-memory effect was first noticed in 1932, when Swedish scientist Arne Olander noted it in a gold-cadmium alloy. At the time it had little practical value; the gold-cadmium alloy was both toxic and expensive. A breakthrough in SME development came in 1963, when researchers with the U.S. Naval Ordnance Laboratory discovered SME in an alloy of nickel and titanium that they were studying for use in marine applications. They called the alloy Nitinol—Ni for nickel, Ti for titanium, and NOL for the Naval Ordnance Laboratory. It was relatively cheap and nontoxic, and had a more pronounced deformation/recovery ratio than any previous SME alloy.

That discovery sparked a surge of research into potential applications of SME alloys, with NASA and defense agencies in the U.S. and overseas leading the way. NASA had explored the technology in the 1960s, and then put it aside until the 1980s, when it began preparation for a space station to orbit the Earth. (The new interest in SME alloys also reflected the advances made since their discovery; scientists and engineers who work on space exploration are constantly experimenting with new materials.) SME alloys were also used to develop satellites that could automatically unfold their antennas when exposed to the heat of the sun.

One of the companies involved in the NASA research was Memry Technologies Inc., a subsidiary of Memry Corporation, Brookfield, Connecticut. In its work for NASA it developed several kinds of memory-metal joints used to connect and disconnect space station structures. It was during this project that the company developed the technology for its own line of memory-metal products. The company began making a line of consumer products to prevent accidental scalding. The devices used a small valve made of an SME alloy that would cut water flow when temperatures rose above 120 degrees Fahrenheit. A second company, Marchon Eyewear, Inc., of Melville, New York uses NASA's memory metal technology in its Flexon eyeglass frames. According to the company the frames can be bent or twisted and still go back to their proper shape. The Marchon frame material doesn't need heating.

SME alloys are typically composed of nickel and titanium (NiTi), or a combination of copper, zinc, and aluminum (CuZnAl). According to *Electronic Design* magazine, shape memory alloys are "trained" through a combination of heat

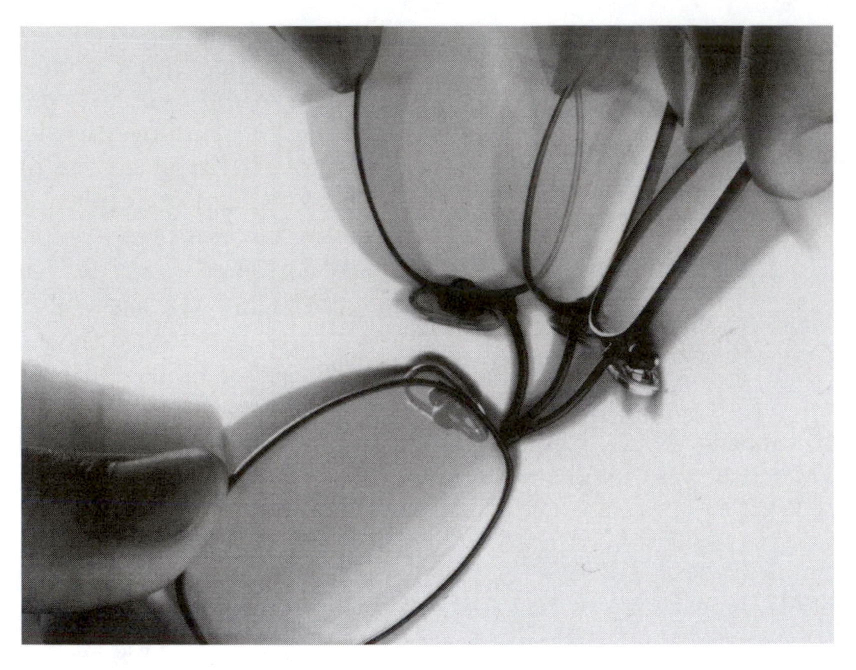

Marchon Eyewear's Flexon frames. Photo courtesy Marchon Eyewear.

and mechanical force. The alloy is first bent to its desired shape, then heated to about 4,000 degrees Centigrade and quickly cooled in water. The sudden cooling freezes the material's crystalline structure in its new position, and the part can't return to its original shape unless heated. Shape-memory alloys can also be trained to have a two-way memory and hold two shapes, changing between the two depending on the level of heat applied.

Shape-memory alloys have been used in a number of other ways—in medical devices, such as tiny filters that open at body temperatures to catch blood clots, or surgical instruments that can be custom-bent during an operation and then later returned to their original shape. The alloys have also been used to make greenhouse windows that open automatically or engine-cooling fans.

Today, the applications of shape memory alloys continue to grow. NASA, too, continues to find new uses for the technology. SME alloys are being used in the development of robots and for new kinds of aircraft. The "Morphing Project" at NASA's Langley Research Center in Virginia is an effort to use smart materials such as SME alloys. Project manager Anna McGowan said in an interview with ABC News that research into next-generation aircraft materials will let NASA "make airplane wings more flexible to maybe fly more like birds, who actually bend and twist their wings in flight."

Additional Reading

Goldberg, Lee. "Computers That Recycle Themselves? Maybe in the Future, with 'Shape-Memory' Materials." *Electronic Design* 46, no. 20 (1 September 1998): 29.

Marchon Web site. www.marchon.com.
"Memory Metals." *NASA Spinoffs: 30 Year Commemorative Edition* (1992): 82.
Pelton, Alan. "The Use of Nickel-Titanium Alloys in Medicine." *Medical Plastics and Biomaterials,* March 1997, 30.

Microthermal Fabrics

Imagine that your clothes can sense when your body is cooling down and that the material itself releases body warmth, stored up from earlier in the day. That's essentially what the phase change materials (PCMs) from Outlast Technologies in Boulder, Colorado, can do. Such microthermal fibers were first developed for NASA and the U.S. Air Force for use in gloves to keep pilots' and astronauts' hands warm.

A North Carolina company, Triangle Research and Development Corporation, had developed these microthermal fibers in 1990. The space agency was in the midst of a financial belt-tightening, so the firm made the technology commercially available. Gateway Technologies, now known as Outlast Technologies, has incorporated it into outerwear, such as jackets and pants, underwear, socks and most recently, mattresses and bedding products.

These fibers do not provide thermal insulation, but a kind of thermal regulation, the company explains on its Web site. It's called phase change material because it refers to the transition from one phase to another—liquid turning to solid, for example. Outlast phase change materials are substances that are calibrated to stabilize in a slush state (half liquid, half solid), the company explains.

Outlast fibers contain microcapsules filled with PCMs inside the fiber strands. These fibers can be woven or knit into other fabrics for use in thermal underwear, socks, gloveliners, blankets, hats, and shirts.

Outlast PCMs can also be applied to the outside of fabrics, such as jackets, gloves, hats, boots and shoes. Outlast technology can also be applied to foam, for use in ski helmets, ski boots, footwear and furniture upholstery.

The problem with ordinary outdoor clothing is obvious. You're hiking on a cold day and your feet get sweaty. When you stop to rest, and your feet cool down, the sweat in your socks can begin to feel icy cold on your feet.

But with Outlast fibers, the "slush" absorbs the excess heat generated by your body. This accomplishes two things: "This keeps you cooler and extends the period of time before your body's own cooling system (sweating) starts in," the Web site states. Then as your body cools down, the phase change materials begin to change back from a liquid to solid. In other words, the excess heat absorbed earlier is returned to you. The company claims the Outlast materials can keep your body temperatures near to normal for the duration of wear. The melting range of the phase change materials can also be adjusted, meaning that material used in firefighters' clothing would react very differently from those of commercial fishermen or aircraft pilots.

Mattress-maker Serta incorporated the Outlast technology into its luxury line of mattresses—Perfect Night. According to the Serta Web site, "Every Perfect Night model is quilted with Sensifiber, an exclusive and revolutionary fiber layer that uses temperature regulation to maintain a comfortable sleeping surface. As it responds to your body temperature, Sensifiber wicks away heat when you're hot and releases heat when you're cold."

Outdoor clothing makers are incorporating the Outlast fibers into their clothing, typically at a 10 to 15 percent increase in cost. Columbia Sportswear Co. designed a parka with Outlast, particularly with the skier in mind. Skiers can find themselves on the mountaintop, chilled from the chair lift ride up. Then as they're skiing down the mountain, they become uncomfortably warm. Outlast technology can also be found in clothes by Marmot, Nike ACG, Wigwam Socks, and Swiss Army Clothing.

In April 2002, Outlast unveiled a line of bedding products, including sleep pads, pillows, comforters and duvets. Company director of sales and marketing for home furnishings, Guy Eckert, told *Home Textiles Today* magazine that, "Outlast eliminates those peaks and valleys and will give the consumer more of a consistent environment as they sleep."

Additional Reading

Cowan, Kevin. "Cold Weather Chic." *Detroit Free Press/Scripps Howard News Service*, 7 December 2001.

Dalton, Greg. "A Shirt That Thinks." *The Industry Standard* 4, no. 24 (June 2001): 52.

Lazaro, Marvin. "Outlast Branches Out into Bedding." *Home Textiles Today* 23, no. 16 (17 December 2001): 4.

Marston, Wendy. "Wonder Wear: Can We Interest You in a Suit That Banishes Dirt, Sweat, and Germs, Sir?" *Discover,* January 2000, 46–48.

McEvoy, Christopher. "Get Smart" *Sporting Goods Business* 30, no. 15 (22 September 1997): 35.

Rastelli, Linda. "Business Company Hopes to Outlast Others: Maker of 'Smart' Sports Apparel Plans New Products." *Denver Post,* 5 February 2001.

"Thermal Clothing." *NASA Spinoff 1997* (1997): 78.

Walzer, Emily. "The Inside Story: With Technical Innovations and Heightened Marketing Campaigns, Insulations Are More Than 'Just Stuffing.'" *Sporting Goods Business* 34, no. 5 (9 March 2001): 32.

Osteoporosis Detection and Prevention

Here on Earth, we take gravity for granted. Space scientists have been trying to mimic some of gravity's beneficial effects in the zero-gravity environment of space for decades. Without gravity, unusual things begin to happen to astronauts' bodies.

The absence of gravity actually causes the shape and size of bone and muscle cells to change. For example, muscles, which no longer fight gravity, begin to atrophy. Muscles can lose about 20 percent of their mass if they're not used. And it happens quickly—at a rate of up to 5 percent a week. Weight-bearing bones, such as legs and the spine, also atrophy. Research has shone that, in space, bones atrophy at a rate of about 1 percent a month and could reach as high as 40 to 60 percent.

The "disuse osteoporosis" that sets in is physiologically similar to osteoporosis affecting millions of older people, primarily post-menopausal women. Literally, osteoporosis means "porous bones." As bones become weaker, they are more susceptible to fractures. While it's generally considered an older person's disease, younger people can develop it as well. Factors such as diet, exercise, hormones and genetics all play a role. According to the American Medical Women's Association, "osteoporosis leads to 1.5 million fractures, or breaks, per year, mostly in the hip, spine and wrist, and costs $14 billion annually. One in two women over the age of 50 will suffer an osteoporosis-related fracture."

Ever since NASA engineers realized that extended time in a microgravity environment had this effect on bones, they've been working on ways to easily measure bone density loss. Several of the tools that were developed have been turned into commercial products. For example, NASA needed a way to take direct measurements of bone stiffness and mass. The result was the development in 1989 of the Mechanical Response Tissue Analyzer (MRTA), a non-invasive instrument developed by a three-way collaboration among Ames Research Center, Stanford University, and Gait Scan Inc. of Ridgewood, New Jersey.

In a February 28, 1995, NASA press release introducing the instrument, it explains that MRTA is a portable device, delivering no radiation, which measures the response of the long bones (either the ulna in the forearm or the tibia in the leg) to a five-second vibratory stimulus. By measuring the bending stiffness of these bones, the MRTA provides a quick analysis of overall bone strength. Not only could the instrument be used on astronauts following flights, but also on patients undergoing treatment for osteoporosis or those who are recovering from broken bones.

More recently, scientists were designing a compact machine to allow bone and tissue measurements to be taken in space. Scientists at the National Space Biomedical Research Institute (NSBRI) are working on the advanced multiple projection dual-energy X-ray absorptiometer (AMPDXA). In a NASA Space Research Briefs article, Dr. Harry K. Charles, NSBRI technology development associate team leader and assistant department head for engineering at The Johns Hopkins University Applied Physics Laboratory, said, "Knowing these measurements while in space will allow astronauts to either increase exercise or take medications to counter the loss of bone and muscle mass due to long-duration microgravity exposures." The tool could be used for diagnosing and monitoring osteoporosis and a portable version could be used in osteoporosis screenings in retirement communities and nursing homes.

On Earth, osteoporosis is generally preventable through a diet with adequate calcium and vitamin D intake and through weight-bearing exercises, such as jogging, tennis, bicycling and weight training. These are activities that combine movement with stress or resistance on the limbs.

But in a zero gravity environment, that resistance is extremely difficult to create. For long-term space travel, bone atrophy would be a serious condition that would require a long rehabilitation period when astronauts return to Earth. So for years, researchers have been working to design exercise contraptions that simulate the resistive forces of gravity. For example, the Russians tried strapping jogging cosmonauts to a treadmill with bungee cords. NASA developed the Interim Resistive Exercise Device or IRED, a system of canisters that can provide more than 300 pounds of resistance for a variety of exercises. A former NASA researcher developed a Lower Body Negative Pressure (LBNP) device—a treadmill inside a chamber that uses the suction of a vacuum cleaner to apply pressure.

Most recently, NASA-funded scientists have come up with a whole new idea: vibrations could prevent bone loss. A November 2, 2001, article on the Science@ NASA Web site details the studies, which have not yet been tested in space. The experiments have put animals—turkeys and rats—on a plate that vibrates slightly for 10 to 20 minutes per day. In the NASA article, researcher Clinton Rubin, a professor of biomedical engineering at State University of New York Stony Brook, said, "Our hypothesis is that a key regulator of bone mass and morphology are the mechanical stimuli that come out of muscle contractions. So instead of these big, intensive deformations of bone, it's basically lots and lots of little ones [that provide a major stimulus for bone growth]."

In other words, by mimicking these twitches, bone mass loss might be prevented. Vibration therapy has entered early clinical trials involving sixty post-menopausal women and also with adolescent girls with very low bone density and children with cerebral palsy. In the studies involving the post-menopausal women, early results are encouraging but preliminary, Rubin said in the NASA article. A broader Phase III clinical trial is being organized. There has also been talk about trying the experiments in space, although that hasn't yet been scheduled. In space, the astronaut could be attached to the vibrating plate and could continue to do other activities. The vibrating movements are barely perceptible.

Actually, gravity does exist in space. In fact, in the region 350 km above Earth's surface, where shuttles and the space station orbit, the pull of gravity is about 90 percent as strong as it is on Earth. But astronauts feel weightless, explains the NASA story, "Space Bones" because "they and their spacecraft are freely falling together toward the planet below. Just as gravity seems briefly suspended in a downward-accelerating elevator, so does the crew in the freely-falling space station experience 'zero-G.' In fact, it's not exactly zero—but near enough. The acceleration they feel is as little as 0.001% of the gravitational acceleration on Earth's surface."

Space is clearly a unique place to study bone loss, but the overlap between astronauts' disuse osteoporosis and the disease that affects millions of people is large. Better measurement systems and any efforts to counteract loss of bone density will help astronauts and the Earth-bound, alike.

Additional Reading

Barry, Patrick L. "Good Vibrations: A New Treatment under Study by NASA-Funded Doctors Could Reverse Bone Loss Experienced by Astronauts in Space." Science@NASA Web site. 2 November 2001. http://science.nasa.gov/headlines/y2001/ast02nov_1.htm. Accessed 8 June 2003.

"Bone Function." National Space Biomedical Research Institute Web site. http://www.nsbri.org/HumanPhysSpace/focus6/ep_function.html. Accessed 8 June 2003.

"A Boon for Bone Research." NASA *Spinoff 1996* (1996): 24.

" 'Good Vibrations' May Prevent Bone Loss in Space." NASA press release, 1 October 2001.

Hullander, Doug and Patrick L. Barry. "Space Bones." Science@NASA Web site. 1 October 2001. http://science.nasa.gov/headlines/y2001/ast01oct_1.htm. Accessed 8 June 2003.

"Measuring Bone Loss in Space and on Earth." *Space Research Briefs,* September/October 2001.

Miller, Karen. "Gravity Hurts (so Good)." Science@NASA Web site. 2 August 2001. http://science.nasa.gov/headlines/y2001/ast02aug_1.htm. Accessed 8 June 2003.

"New Medical Instrument Developed." NASA press release, 28 February 1995.

"Q&A: Osteoporosis." American Women's Medical Association fact sheet. 1999. http://www.amwa-doc.org/healthtopics/osteoporosis_qa.htm. Accessed 8 June 2003.

Pacemaker Improvements

Thousands of people owe their very lives to NASA technology. NASA's collaborations have been instrumental in some of the major advances in heart pacemakers, devices that generate electrical pulses to the heart to control heart rhythms.

In the 1970s, NASA, the Johns Hopkins Applied Physics Laboratory, and Pacesetter Systems, now part of St. Jude Medical, brought about several developments. The first development was a rechargeable, long-life pacemaker battery. NASA had created similar technology for the electrical power systems on spacecraft. Another development was the first single-chip pacemaker, which used miniaturization technology developed by NASA researchers. More recently, NASA's technology for two-way communications satellites became the foundation for implanted pacemakers that can be reprogrammed without requiring additional surgery.

Instead, the pacemaker communicates through this wireless telemetry to a physician's computer console. The pacemaker collects data about the patient's heart rhythms and can be programmed to generate the pulses as needed. That data can be read by the physician, who can see when a patient's heart is beating abnormally. When the device senses irregular heartbeats, it automatically delivers an electrical stimulus to correct the rhythm. Because doctors can reprogram the pacemaker outside the body, making adjustments is faster and much less traumatic than corrective surgery. With pacemakers, patients who have irregular heartbeats can lead healthy active lives.

These pacemakers also contain different pacing functions. In other words, when a person is swimming or jogging, the heart rate quickens to pump more oxygen-rich blood to muscles. A pacemaker can stimulate a quicker heart rate.

Since being introduced in 1979, the technology has become more refined and much smaller. St. Jude Medical now makes the Microny® II SR+AutoCapture™ Pacing System, a pediatric pacemaker about the size of two quarters stacked on top of each other. In June 2001, a three-month-old boy became the first patient to receive the new pacemaker. It's specifically designed to treat patients with bradycardia, or potentially debilitating slow heart rates.

This particular pacemaker is about 40 percent of the size of previous pacemakers. It weighs only 12.8 grams, compared to other pediatric pacemakers that weigh about 18 grams.

Less than a year after it was first used, in March 2002, surgeons implanted the pacemaker in a baby girl who was born eleven weeks premature and weighed less than three pounds. Both of these babies will likely require pacemakers for their entire lives. As they grow, larger pacemakers will be implanted.

Pacemakers are comprised of two parts: one is the component that actually generates the electrical impulses and collects the data. The other part is lead wires that are connected directly to the muscles of the heart. In very young pediatric patients, the pacemaker generator is usually placed in the abdomen near the heart; the more traditional place is by the shoulder.

The bidirectional telemetry that creates the communication signals used in programmable pacemakers has also been part of a vastly different spin-off—in racecars. Formula F1 race cars will have tiny aerials mounted on the side of the cars that will send data back to the computers in the pits. Crew engineers can analyze the data and make quick adjustments.

See also:

Implantable Cardiac Defibrillator, Heart Pump, Heart Rate Monitor.

Additional Reading

"Advanced Pacemaker." *NASA Spinoffs: 30 Year Commemorative Edition* (1992): 32.
Dunbar, Brian. "Space Technology Used to Detect and Treat Heart Disease." NASA Cardiac Research news release, 20 February 1998.
"LLU Medical Center Implants World's Smallest Pacemaker in First United States Patient." Loma Linda University press release, 7 June 2001.
Maurer, Allan. "Minicourse: They Came from Outer Space." *Modern Maturity,* December 1988/January 1989, 57–61.
"Programmable Pacemaker." *NASA Spinoff 1996* (1996): 25.
"VCU Surgeon Implants Pacemaker in Premature Baby." Virginia Commonwealth University press release, 8 March 2002.

Pill Transmitters

When astronauts are in space, their vital statistics are being monitored on Earth. In the early years of space travel, that was accomplished by putting sensors on the astronaut's body. The sensors were connected by wire to the receiver. The problem, of course, was that the wires got in the way.

The next step was to gather the data by telemetry. In this instance, data could be transmitted to the receiver without wires. Astronauts were free to move about and do whatever tasks they needed to. In order to make the monitoring process even more unobtrusive, researchers from the Sensors 2000 team at NASA's Ames Research Center have developed the Implantable Telemetry System. This transmitter fits inside a plastic pill-shaped container placed in the body. It can measure heartbeat, blood pressure, or pH level and send that information to the receiver. The pill itself is about one-third of an inch wide by just over an inch long. Although it is called a pill, it is not swallowed, but actually implanted in the body.

It hasn't yet been used with astronauts, as researchers are working on a smaller version that astronauts can swallow prior to a mission. But researchers working with physicians at the Fetal Treatment Center at the University of California, San Francisco, have found another extraordinary use for the tiny transmitter: as a fetal monitor for at-risk babies.

In 1981 the first corrective surgery was performed on a fetus. Since then, surgical procedures to correct several life-threatening defects have been improved upon. While the surgeries are often successful, they raise a new concern. The risk of pre-term labor is higher after fetal surgery. It is often difficult for women to tell if they have entered into pre-term labor, but the stress it places on the fetus is dangerous. So starting in 1991, surgeons at the Fetal Treatment Center started implanting large transmitters that would send some essential data to a receiver, and in that way they could monitor pre-term labor, if it began.

However, the larger transmitters that were being used were now too large. As the field of fetal surgery improved, incisions became smaller and surgeons used tiny endoscopic tools. Recovery from small-incision, less-invasive surgeries is typically much faster. These pill transmitters can be implanted through three endoscopic incisions. Powered by two small batteries, the transmitter works for four to six months. It measures the magnitude and frequency of contractions, allowing doctors to spot pre-term labor before it becomes life threatening to the fetus.

The transmitters would allow a pregnant woman recovering from surgery to return home rather than have to stay in the hospital.

Also in the research and development stages are even smaller transmitters that could be implanted into the fetus itself. These transmitters would supply even more information, such as fetal heart rate. Researchers are also working to make these sensors flat and attach them to a bandage-like adhesive that could be placed on the body.

The pill transmitter also has applications in farming, according to NASA. In its NASAExplores section of the Web site, NASA researchers explain that "Cows have a nasty habit of swallowing bits of barbed wire fencing and tin cans while grazing on grass and wildflowers. If that metal gets into a cow's milk, it can make it sour. Right now, instead of telemeters, farmers rely on magnets. Cows have a series of four stomach areas, called rumen, and three-inch magnets are fed to the cows. These magnets stay in the third rumen to attract and trap bits of metal. The transmitter could be attached to the magnet and would detect changes in the cow's pH level that might result in sour milk. Farmers can then take action."

The number of applications for this kind of technology is sure to increase. Researchers already see possibilities for using pills to monitor intestinal pressure changes or stomach acidity in ulcer patients.

Additional Reading

"Eavesdropping on Baby" *NASAExplores,* 8 March 2001. http://nasaexplores.com/lessons/01-013/5-8_index.html. Accessed 8 June 2003.

Fetal Treatment Center. www.fetus.ucsf.edu.

Jennings R. J., N. S. Adzick, M. T. Longaker, and M. R. Harrison. "Radiotelemetric Fetal Monitoring during and after Open Fetal Surgery." *Surgical Obstetrics & Gynecology* 176 (1993): 59–64.

"Miniaturized Transmitter to Be Used in Efforts to Save Babies." NASA press release, 18 November 1998.

"Miniaturized Transmitter to Monitor At-Risk Pregnancies." *Medical Equipment Designer* (January 1999).

Philipkoski, Kristen. "Womb with a View," 8 December 1998. http://www.wired.com/news/technology/0,1282,16703,00.html. Accessed 8 June 2003.

"Space Age Sensor Helps Save Infants' Lives." NASA press release, 15 August 1995.

Pool Filters

There are no swimming pools in space and there never will be. But there is, inside the confines of a spacecraft, plenty of water. It's there for the crew to drink, bathe in, and to handle waste materials. Keeping the water supply clean is essential for the crew's health. In the 1960s and early 1970s during the missions of the Apollo program, engineers at the Johnson Space Center in Texas began developing a new kind of water filter for long trips in space—one that was lightweight, small, required little power to operate, and that astronauts wouldn't have to monitor.

The system they developed was only slightly larger than a pack of cigarettes and weighed just nine ounces. It uses the microbe-killing power of silver, a metal long known for its ability to control bacteria, in a process known as ionization. When a small electric current travels between two silver electrodes, small levels of metal ions break loose from the electrodes and enter the water supply, where they prevent the growth of bacteria. By placing the ionization devices at various spots in the spacecraft's water and wastewater supply systems, scientists were able to release silver ion concentrations of between 100 and 300 parts per billion, enough to kill bacteria in the water within a few hours, according to NASA's *Spinoff* publication. Besides being an easy and effective way to keep water clean, astronauts no longer had to deal with a system that used chemicals to purify the water.

Swimming pools, of course, are a breeding ground for bacteria as well. A South Carolina company, Caribbean Clear, Inc., used NASA's technology to make a pool filter that could keep water clean without heavy doses of chemicals such as chlorine and bromine. The company, founded in 1983, developed an ionization filter that—like the NASA device—used silver to kill bacteria. It also added copper, a metal known for its ability to kill algae. It has two silver/copper alloy electrodes that release silver and copper ions when an electric current passes through them. Water is pumped past the electrodes, sweeping away many of the ions before they can travel from one electrode to the other. The device also switches the polarity between the two electrodes to prevent plating, the accumulation of a thin layer of metal, which would shorten the system's life. There are also monitors that automatically adjust the release of ions.

Today a number of companies offer ionization devices to clean pool water, as well as hot tubs and fountains. From Clearwater Pool Systems, a Florida company, comes this description of how copper and silver purify water:

> When copper and silver ions are released into the water, these cationic, surface-active ions are a potent biocide. The disinfection action takes place when the positively charged copper and silver ions form electrostatic bonds negatively charged areas on the microorganism cell walls. These electrostatic bonds create stresses, which lead to distorted cell wall permeability, minimizing the normal intake of life sustaining nutrients. Once inside an algae cell, copper and silver attack sulfur containing amino acid residues in the proteins used for photosynthesis (the process of converting light into usable food and energy). As a result, photosynthesis is blocked and leads to cell lysis and death. If the algae cell manages to live, the reproduction process is hampered by the presence of the copper ions and the spread of algae is held in check. Bacteria is killed, rather than suppressed.

According to the companies that manufacture these devices, ionization offers advantages over using chemicals such as chlorine, which is harsh and capable of irritating skin and eyes. With ion generators there is less need for chemicals because the devices can do much of the work of keeping bacteria levels at safe levels. Plus, it's cheaper than buying chemicals every summer.

In general, according to pool industry publications, the standards for pool water mirror the Environmental Protection Agency standards for drinking water (although even pool water that meets those standards isn't considered safe to drink). EPA guidelines currently call for 1.3 parts per million for copper, and 0.1 parts per million for silver.

In fact, even though the technology for using ion generators to clean water is relatively new, knowledge of the purifying effect of silver and copper in water has been known for centuries. The ancient Greeks discovered the power of copper to prevent algae growth, and early Egyptians kept their water in silver containers to prevent contamination.

Additional Reading

Caribbean Clear Inc. www.caribbeanclear.com.

National Sanitation Foundation International. www.nsf.org/pools.

Palmer, Brian. "Just Add Water: Innovative Designs for Backyard Swimming Pools." *New York Times Magazine,* 18 July 1999, 18.

"Pool Purification. Space-Based System Yields Pool and Spa Water That Exceeds EPA Standards for Drinking Water." *NASA Spinoffs: 30 Year Commemorative Edition* (1992): 88.

Quantum Tubers

Potatoes are one of the world's most important food crops. And NASA technology, along with research done by the Chinese, is making it easier and faster to bring new healthy potato seed to market.

Researchers are always searching for ways to make plants more productive. For example, potatoes that are more drought-resistant could be planted in areas where rainfall is typically scarce. Potato plants that produce more potatoes increase the overall harvest. But bringing a new potato to market could take years. Here's why.

First, it takes the researchers several plantings to develop and test the new potato. Once they're sure they have a healthy potato plant, they need to create seeds or very small potatoes that are sent to farmers for planting. To create enough seed stock usually takes five to seven field plantings in most climates or even in greenhouses—that translates into about five to seven years.

By the early 1990s, Chinese researchers had figured out a way to speed up the process considerably by taking tissue cuttings from the new plants and growing them. By doing this in a lab, they were able to create thousands of genetically identical seedlings. The nine-month process of growing new potatoes in a field was reduced to forty days in a lab. However, growing enough plants in a greenhouse was labor-intensive.

Here's where Americans and the space agency come in. Researchers at the University of Wisconsin-Madison heard about what the Chinese were doing and contacted Robert Britt, president of American Ag-Tec, a biomanufacturing company based in Delevan, Wisconsin. Together, they teamed up to repeat the Chinese experiments at the university's Biotron—a biological research building. The Biotron allows researchers to manipulate the physical environment in one of its 50 rooms with artificial lighting or 29 greenhouse units. Being able to control the experiments more carefully, the researchers doubled the success of the Chinese experiments.

The group knew that NASA was also conducting some research on growing potatoes at Biotron. The NASA-sponsored Wisconsin Center for Space Automation and Robotics (WCSAR) was testing a growth chamber using Light Emitting Diodes (LED) technology, high-efficiency temperature, and humidity controls to grow potatoes in a closed facility without labor-intensive handling. Minituber growth was flight-tested in October 1995 during a launch of the Space Shuttle *Columbia* in its Microgravity Astroculture Laboratory, and they were delivered for testing aboard the Russian space station *Mir* in January 1998.

The experiments were a perfect fit because the NASA technology allowed American Ag-Tec to bring the production of these minitubers indoors where they could be grown year-round. A Quantum Tuber factory could turn out 10 to 20 million tubers each year.

Plus the field life cycle is now down to two plantings. In other words, the entire process of getting healthy, pathogen-free seed stock is now five years shorter than it

Quantum Tubers. Photo courtesy NASA.

had been. Bringing new seed stock to market is important because farmers who continue to use old potatoes as next year's seed stock end up with poorer quality and disease-prone potato crops.

The first Quantum Tuber production began in Poland in 1999. As an article in the Chicago-area *Polish Suburban News* explained, "Polish potato growers desperately need new varieties of plants. It is estimated that about 85 percent of the potatoes in Poland are so diseased that they qualify only for livestock feed."

Because the system can be used for growing regular potatoes as well as genetically modified potatoes, it may also contribute to another developing area: edible vaccines. Researchers in the United States and other countries are working on producing genetically modified plants that make antibodies against hepatitis B—a liver infection that affects two billion people worldwide. Other work is ongoing in creating plants that generate antibodies to cholera, bacterial tooth decay, lung infections in patients with cystic fibrosis, AIDS, and sexually transmitted diseases.

American Ag-Tec was contracted to grow potatoes containing a hepatitis B vaccine by Axis Genetics, a British company testing the concept in mice and human trials. Creating edible vaccines would be far less expensive—and less painful—than the injected versions.

Research is also ongoing to genetically modify plants to protect livestock from various illnesses.

Additional Reading

Bergquist, Lee. "Company to Grow Potatoes for Vaccine." *Milwaukee Journal Sentinel*, 1 January 1999.

Boyd, Vicky. "Technology Mass Produces Minitubers." *The Grower,* January 2000.
Mika, Elizabeth. "Space Spuds and Edible Vaccines." *Polish Suburban News,* July 1999.
Paprcka, Susan. "Biotech Boom: How Wisconsin's Biotech Industry Could Become a World Leader." *Small Business Times,* 28 April 2000.
Quantum Tubers. www.quantumtubers.com.
"NASA Spins Off Revolutionary Seed Potato Technology." Quantum Tubers press release, 17 October 2000.
"Space Age Spuds" *NASA Spinoff 2000* (2000): 55. http://www.sti.nasa.gov/tto/spinoff2000/er5.htm. Accessed 8 June 2003.

Race Car Insulation

When a Space Shuttle reenters the Earth's atmosphere, temperatures can flare as high as 3,000 degrees Fahrenheit. The Thermal Protection System (TPS) tiles and insulating blanket material developed for NASA by Rockwell Space Systems (now part of Boeing) protects the astronauts from these searing temperatures.

In the first commercial use of shuttle TPS technology, similar insulation is being used to protect occupants of another kind of craft hurtling past at high speeds: race cars. Without this kind of insulation, the area below the driver's foot, just above the car's exhaust system, can reach temperatures of 145 degrees Fahrenheit. The exhaust pipes on NASCAR cars are so close to the sheet metal of the floor pan that they radiate a great deal of heat into the cockpit. Experts predict that the overall temperature in the cockpit can approach 160 degrees Fahrenheit. In some spots, it gets even hotter. The heat can rise to 260 degrees Fahrenheit in the area around the driver's left elbow. So even with their fire-retardant suits and forced air-cooling systems, the cockpit of a race car can be dangerously hot. Drivers have received localized second- or third-degree burns during races.

In 1996, NASCAR driver Rusty Wallace tested the TPS insulating blanket at the Daytona International Speedway. He drove 50 high-speed miles with the exterior insulation in place. Temperatures were measured at various points inside the cockpit. Later, he drove the same distance without the insulation. The comparison showed that the TPS material reduced the heat in the car's interior between 30 and 50 degrees Fahrenheit.

At a press briefing after the test runs, Wallace said, "I feel that the TPS material helps the whole car run cooler, and the cooler the car is the better the performance." He also said that overall safety could be improved as a result since drivers will be more alert and have better concentration, according to a report in NASA's publication *Spaceport News.*

The first driver to use the TPS material in an actual race was Hueytown, Alabama, native Doug Reid in the May 1997 Busch Grand National Race at Charlotte, North Carolina. Gauges showed that the temperature at the gas pedal never exceeded 90 degrees Fahrenheit. "For the first time in my racing career I didn't even break a sweat," Reid said in an interview published in a NASA newsletter.

According to *Spaceport News,* "The idea of using TPS materials to insulate against excessive heating in the cockpit of a race car came about as a result of a tour taken by NASCAR champion Bobby Allison at Kennedy Space Center. Former KSC Director Jay Honeycutt recommended that TPS insulation materials could shield drivers from the internal high temperature of race cars. Bobby Allison contacted Roger Penske, who provided a car to successfully test the TPS material."

Race car insulation. Photo courtesy NASA.

Three TPS tiles were first tested on Space Shuttle *Columbia* in March 1996. The tiles, developed by Rockwell and NASA's Ames Research Center, were placed in damage-prone areas on the orbiter's base heat shield and lower body flap.

Since the 1996 test, BSR, in Wilton, Connecticut, changed the materials to make the blankets more durable and less expensive. The nonflammable blanket, made of metal, ceramic and glass, is less than a half-inch thick. It is installed over the top of the exhaust system, leaving the bottom exposed to airflow. That prevents overheating. The blankets are also available for installation under the seat and in the rocker panel, driver side door, and exhaust and oil tanks.

BSR is also introducing the kits into the Sports Car Club of America (SCCA) off-road series and to airplane manufacturers.

Additional Reading

"Rockwell Successfully Flight Tests Advanced Thermal Protection." Boeing press release, 18 March 1996. http://www.boeing.com/space/rss/whatsnew/1996/tps33rl2.htm. Accessed 8 June 2003.

"Shuttle Insulation in Race Cars Gets Seal of Approval." *Spaceport News,* 11 October 1996, 4.

"Space Shuttle Insulation Cools NASCAR Driver." *Aerospace Technology Innovation* 4, no. 3 (July/August 1996).

"Thermal Insulation Protects Drivers" *Aerospace Technology Innovation* 8, no. 2 (March/April 2000).

Railroad Anti-Icer

Ice buildup on the switches of a railroad track is a dangerous thing. In the past, ice removal would be done manually—and often—with anti-freeze liquids or electrical heating components. But NASA technology and research on anti-icing agents has made a difference in rail safety, both on the nation's major rail lines as well as on mass transit systems, which use an electrified third rail. Ice buildup on the third rail results in mass transit slowdowns and frustrated commuters.

NASA's Ames Research Center and Midwest Industrial Supply, Inc. of Canton, Ohio, developed Zero Gravity—a smart fluid technology that is used to enhance the company's anti-icing compounds. Anti-icing compounds with Zero Gravity can be poured or sprayed on like a liquid—before ice develops on the rail lines. The liquid then turns into a resilient gelatin-like coating that stays put, even on vertical surfaces. The anti-icing compounds are activated by freezing rain, sleet, or snow. The anti-icer compounds can be activated days or even weeks after the liquid is first applied.

The anti-icing product is biodegradable, noncorrosive, and doesn't conduct electricity like other liquids do. While it forms a .01-inch thick coating that defies the effects of wind, rain, and gravity, it does not feel sticky or tacky, so it won't gum up rail contact shoes—the part of the train that conducts the electricity from the electrified third rail. It is effective in bitterly cold temperatures as low as –70 degrees Fahrenheit. That's an advantage over rail heaters, which can malfunction in subzero temperatures. Plus, the anti-icing agent doesn't consume the energy that rail heaters do.

The company also developed an application system for the Third Rail Anti-Icer. It's called Ice-Slicer and is a dispensing system and tank of fluid installed in a train car and automatically applies the compound to both sides of the third rail as the train moves along the tracks.

The agreement between NASA and Midwest Industrial Supply also resulted in the Ice Free Switch—for use specifically on railroad switches. As freezing rain or sleet hits the coated track or switch, no ice can form.

In the past, railroad crews would manually deice switches and critical sections of rail as needed. Deicing is done only after the ice has formed. Anti-icing is preventative. But when needed, the Midwest anti-icer can be used on frozen switches as well. According to *NASA Spinoff 2001*, "Manually freeing frozen switches can take an entire crew several hours. With the Ice Free Switch, it takes only five minutes to treat the switch by spraying, brushing, or pouring on the product. Ice Free Switch requires as little as one gallon per switch whereas other deicing fluids require five to ten gallons of liquid to effectively melt ice."

Additional Reading

"Anti-Icing/De-Icing Fluid." *Railway Age* 201, no. 2 (February 2000): 62.

Harvey, Fiona. "NASA Inspires Anti-Freeze Fluid." *The Financial Times,* 11 January 2002, 9.

Higgins, Amy and Sherri Koucky. "Nontoxic, Noncorrosive De-Icer." *Machine Design*, 21 February 2002, 29.

Midwest Industrial Supply Inc. Web site. www.midwestind.com.

"New Fluid Prevents Railway Ice." *NASA Spinoff 2001* (2001): 70–71.

Rescue Cutters

It can take a lot of power to cut through the wreckage following an automobile accident or building collapse. One company harnessed the power used to separate the main rockets from spacecraft to develop a portable emergency rescue cutter.

Earlier rescue cutters used gasoline powered hydraulic pumps, hoses, and cutters. These systems weighed 50 to 100 pounds and took about 10 minutes to set up on the scene of an emergency. Once set up, they weren't easily moved. Hi-Shear Technology Corp. of Torrance, California, decided to apply its pyrotechnics knowledge from working on NASA projects to developing a smaller, yet powerful, cutter.

Hi-Shear was already manufacturing pyrotechnically actuated separation devices—essentially small contained explosions. These cartridges, known as initiators have been used in thrusters, explosive bolts, separation nuts, and pin pullers from Apollo and Saturn missions all the way to the Space Shuttle program.

To develop the emergency cutters, the company asked the City of Torrance Fire Department for help. The result was LifeShear cutters, 8- and 12-pound devices that are ready to use within a minute. Miniaturized versions of the cartridges used in space provide the power of an explosive charge to cut through thick materials.

NASA noted that the cutters not only weighed less than former models, but they also cost much less. When introduced in 1994, the cutters cost less than $2,000. Hydraulic models cost between $15,000 and $40,000.

Because of their small size and comparatively low cost, most fire and rescue departments would be able to afford these cutters. Plus, they're easily transported and light enough to be standard equipment on rescue helicopters.

The cutters gained some national publicity when the company donated four LifeShear cutters and 400 power units to the rescue efforts at the Alfred P. Murrah Federal Building following the Oklahoma City bombing in April 19, 1995. The cutters proved their capabilities of cutting through concrete, steel reinforcement bars, wood, and other debris. The Federal Emergency Management Agency (FEMA) found the portable devices so useful, they ordered nearly 40 additional cutters and 6,500 power units. The attack killed 169 men, women and children. Search efforts continued for two weeks as rescue units looked for survivors.

Additional Reading

Hi-Shear Technology Corp. Web site. www.hstc.com/products/lifeshear.htm.

"Hi-Shear Technology's LifeShear Proves Indispensable in Rescue Efforts in Oklahoma City." Hi-Shear press release, 28 April 1995.

"New Light, Portable Fire Department Cutters Use Space Technology to Speed Rescue." Hi-Shear press release, 13 June 1994.

"Rescue Equipment" *Spinoff 1995* (1995): 62.

Riblets for Drag Reduction

Friction is a real drag. That's why NASA researchers have long been interested in surfaces or surface coatings that can reduce aerodynamic drag. Riblets, or tiny v-shaped, barely noticeable grooves in a surface, seem to do just that. The grooves smooth out air or water turbulence.

It's a concept that has a number of applications. NASA tested riblets on aircraft to see if such coatings could improve fuel efficiency. Riblet coatings have also been used on racing yachts and bathing suits. In tests on a Learjet in 1986, the riblets demonstrated an in-flight drag reduction of about 8 percent. The following year, the winning yacht in the America's Cup, *Stars and Stripes*, had the underside of its hull coated with riblet material made by 3M.

It was during the 1970s oil embargo that NASA began searching for ways to increase fuel efficiency by decreasing drag.

Riblet swimsuits. Photo courtesy NASA.

In wind tunnel tests at NASA's Langley Research Center in Hampton, Virginia, researchers worked with models that had riblets inscribed onto their surface. But that wouldn't work for something as large as an airplane. The results from these tests were published in a technical briefing and 3M Co. developed a way to inscribe riblets on a flexible adhesive film that could be used on automobiles and aircraft.

According to the 3M Web site (www.3M.com), to create the shark skin coating on the hull of *Stars and Stripes*, the company used microreplication or the science of designing tiny, three-dimensional structures and reproducing them onto an adhesive film.

By the mid-1990s, the design had been incorporated into bathing suits. Arena North America created the Strush SR swimsuit, which used a silicon ribbing in the chest and buttocks area of the swimsuit—where swimmers encounter the greatest amount of drag. Tests in water flumes showed the suits were 10 to 15 percent faster than other competition suits. The riblets in a suit for a breaststroker would be different from those of a backstroker, for example, because the resistance is different. The suits made quite a splash in 1995 at the Pan American Games as swimmers wearing Strush SR suits won thirteen gold medals, three silver, and one bronze.

Research continues on how surface coatings can make swimmers, boats, and airplanes faster. For example, in swimsuits, the more recent trend has been toward suits covering more of the swimmer's body. The less skin and body hair exposed, the lower the friction. Today's suits also repel water better than ordinary nylon suits, which absorb water, increasing drag. Other competitive athletes are also looking at similar designs. Runners, speed skaters, and bicyclists could benefit from reduced aerodynamic drag.

The airline industry is also continuing similar research. According to a 1997 article in *Design News,* researchers were very pleased with surface coatings that used a series of random bumps to reduce drag. These random bump patterns reduced skin friction drag between 12 and 13 percent. "Some estimate that a typical commercial aircraft can save between $100,000 and $120,000 a year in fuel costs by cutting overall drag just 1%. Worldwide, a 1% drag reduction could translate to fuel savings of more than $1 billion a year."

Additional Reading

Arena Web site. www.teamarena.com.

De Gaspari, John. "America's Cup Spurs Product Development" *Boating Industry* 50 (August 1987): 76.
"Fly like a Shark." *The Economist* 307, no. 7548 (30 April 1988): 86.
Letcher Jr., John S., John K. Marshall, James C. Oliver III, and Nils Salvesen. "Stars & Stripes: Computer Aided Design of America's Cup Winner." *Scientific American* 257 (August 1987): 34.
Minerd, Jeff. "Shhh! Engineers at Work on Noise." *The Futurist* 33, no. 8 (October 1999): 8.
Murray, Charles J. "From Aircraft to Ship Design: Get Ready for a 'Bumpy' Ride." *Design News* 52, no. 25 (15 December 1997): 25.
Rosenberg, Barry. "Speed in the Groove." *Technology Review* 90 no. 8 (November–December 1987): 10.
Vaughan, Christopher. "Saving Fuel in Flight: Projects Conceived in the Oil-Poor 1970s Now Bear Fruit." *Science News* 134, no. 17 (22 October 1988): 266.
Walsh, Julie. "Suits Slicker Than Skin." *Swimming World,* July 1998.

Robotic Surgery

Space is the perfect place for robots. Using machines in place of human beings to do work in the harsh environment of space holds great promise for the future of space exploration. Besides the obvious benefits of allowing astronauts to stay in the safety of their spacecraft, imagine how much more work could be done by robotic equipment that never needs food or rest, for example, on board the International Space Station or with a Space Shuttle payload. Plus, robotic devices like the Mars Surveyor can explore places humans cannot reach.

NASA has set a goal of having half of the tasks performed outside its spaceships done by remotely controlled robotic equipment by the year 2004. Engineers working with Cal Tech's Jet Propulsion Laboratory in California have already made advances in designing sophisticated robotic equipment in anticipation of future missions to space. Those advances are being used today to outperform humans on Earth as well. A robotic surgical tool called ZEUS, designed by the California-based company Computer Motion, is able to do what human surgeons cannot. The ZEUS system lets surgeons performing endoscopic surgeries—which use a slender camera inserted into the patient to show the part of the body being operated on—to work with far greater precision.

Endoscopic surgeries are already common, and they're valued because they require incisions of only a few centimeters long instead of the much larger incisions needed for surgeons who operate by hand. But endoscopic surgery still requires significant training and dexterity. Robotic surgery, though, can scale down the movements of a surgeon's hands. A doctor seated before a computer screen just a few feet from the patient and controlling a robotic arm can make natural hand movements that are translated into precise motions.

That control is invaluable for delicate surgeries on areas such as the eyes, brain, or spinal cord. For years physicians have been able to use microscopes to see tiny lesions or tumors in the body, but operating on them required skills that took years to develop. NASA robotic technologies now allow surgeons to not only make precise movements, they also make it possible to use microinstruments—tiny surgical tools—to perform operations that would have been impossible with human hands.

In collaboration with NASA's Jet Propulsion Lab, a New Mexico-based company called MicroDexterity Systems has developed a robotic microsurgery

The Zeus Robotic Surgery System. Photo courtesy NASA.

device called the Robotic Assisted Micro-Surgery (RAMS) workstation. The surgeon moves a single joystick-like handle, and those motions are translated by computer. Robotic surgery equipment can handle complex tasks, such as tying knots with suture material finer than a human hair.

The RAMS workstation makes it possible for surgeons to accurately place surgical tools to within 20 millionths of a meter, and to scale down their hand movements by as much as five to ten times, along with eliminating involuntary jerks and hand tremors. Potential NASA applications include extra- and intravehicular activity telescience, bioprocessing, material process assembly, and micromechanical assembly. It may also have Space Station-related applications in biomedicine.

Perhaps the best-known robot in NASA's program is the *Mars Pathfinder*, which landed on the red planet in 1997. Scientists were able to remotely control the movements of the six-wheeled solar-powered rover called Sojourner, guided by images the rover transmitted back to Earth. A similar kind of telerobotics is happening in medicine as well. Using robotic surgery equipment surgeons are now able to operate on patients thousands of miles away. In September 2001 two doctors in New York City performed gallbladder surgery on a 68-year-old patient 3,900 miles away in Strasbourg, France. The 54-minute surgery was the first complete transatlantic remote control surgery. It showed the potential the technology has to let surgeons work on patients in remote areas, on battlefields, and in space. It is not just a new tool for medicine. Telerobotics has applications in other fields as well, such as maintenance of nuclear power facilities or any potentially hazardous work area. Plus, the ability of robots to perform complex tasks better and faster than human beings opens up a wide range of possible applications.

Additional Reading

"French Connection: With a Transatlantic Operation, Two Doctors Define the Cutting Edge in Robotic Surgery." *People Weekly*, 8 October 2001, 103.

Mack, Michael J. "Minimally Invasive and Robotic Surgery." *The Journal of the American Medical Association* 285, no. 5 (7 February 2001): 568.

MicroDexterity Systems Web site. www.microdexsys.com.

Salisbury Jr., Kenneth J. "The Heart of Microsurgery." *Mechanical Engineering-CIME* 120, no. 12 (December 1998): 46.

Satellite Radio

Have you ever been in a car, straining to hear a favorite song on the radio through annoying static? Space technology could make that a problem of the past.

High above the east and west coasts of the U.S., 22,000 miles up, there are two satellites in a geostationary orbit, named Rock and Roll. Geostationary means they stay fixed above the same point on Earth. They're the satellites beaming 100 channels of continuous, commercial-free radio programming to subscribers of XM Satellite Radio. Sirius Satellite Radio has three of its own satellites delivering another 100 channels of commercial-free programming throughout the continental United States. In 1997, the Federal Communications Commission (FCC) awarded XM and Sirius specific frequencies to be used for satellite radio services.

What makes satellite radio of such high quality is that the system uses digital, rather than analog signals. The result is a CD-like sound, rather than a crackly reception or one that fades in and out.

NASA's involvement in this use of space technology dates back to the late 1980s when it began working with Voice of America, which was updating its high-frequency transmission system. In 1993, NASA began providing the lab space and technical people to support and maintain Digital Audio Radios (DAR) receiver systems at its Lewis Research Center. The Electronic Industries Association (EIA) actually conducted the tests of seven kinds of receivers.

Only XM and Sirius transmit their broadcasts to the continental United States. WorldSpace transmits to Africa and Asia via its satellite radio programming.

XM Radio uses two Boeing HS 702 satellites. "Rock" was launched in March 2001 and "Roll" followed two months later. By the end of March 2002, XM had 76,000 total subscribers—48,000 of them had come on in the previous quarter.

Sirius has launched three SS/L-1300 satellites that orbit in an elliptical path. Each satellite is over the continental United States for about 16 hours a day and at least one of the satellites is over the United States at all times. Both companies have an additional satellite on the ground, ready to be launched in case there are problems with one in orbit.

In each case, the radio company creates original programming or arranges with other broadcasters, such as CNN, to provide content. The content is beamed to the satellite that transmits it back to digital receivers in cars and homes. XM is focusing on both markets; Sirius is primarily promoting car radio programming.

In urban areas where reception might not be as good because of tall buildings, the companies have added ground antennas or repeaters, which ensure a strong signal. The result is that anyone who has a digital receiver and a subscription can tune into the satellite radio. Most people are familiar with the concept as it's used in satellite TV.

One of the questions that remains is whether people will be interested in paying for radio programming. It has traditionally been "free"—paid for by the advertisers who run commercials during programs. But satellite radio is a subscription service. In mid-2002, XM was charging $9.99 a month and Sirius was charging $12.99 a month, plus the price of a digital radio receiver.

Major radio manufacturers got on board, producing digital home and car radios that can navigate the satellite channels as well as display song titles, artist names and other information. Sony, for example, makes an XM radio receiver that can be used in the car and adapted for use in the house.

Sirius, which is concentrating on the car radio market, makes adapters so people with regular car radios can receive the satellite signal.

The companies are competing for subscribers as well as for placement in new cars. XM-equipped audio systems became options in the 2002 Cadillac DeVille and Seville models and additional car makers were planning on introducing the radio in their cars later that year. Cadillac also announced it would become the first car maker to offer XM Satellite Radio on all its models in 2003. Parent company General Motors began offering XM on 25 car and truck models in its 2003 lineup. Chrysler Group announced it will be adding a Sirius Satellite Radio as a dealer-installed product on a line of 2003 vehicles. Most major car and truck manufacturers have said they'll be adding the feature in the next few years.

Satellite radio technology was inducted into the Space Technology Hall of Fame in 2002. The Space Foundation release announcing new inductees states, "Proponents say this could do for radio what cable and satellite did for TV, namely expand programming and raise the quality of an entertainment and communications medium that essentially hasn't changed in decades."

Additional Reading

"Cadillac First Automotive Brand to Offer XM Satellite Radio on All 2003 Models." XM press release, 10 April 2002.

Chase, Victor D. "CD Sound from Space and on the Ground." *Appliance Manufacturer* 42, no. 10 (October 1994): 8.

"Chrysler Group Announces Vehicle Line-Up and Pricing for Sirius Satellite Radio." Sirius press release, 12 June 2002.

Gordinier, Jeff. "Radio Heads: Two Satellite Companies Are Waging a Multi-Billion-Dollar Space War..." *Entertainment Weekly*, 7 June 2002, 28.

Sirius Satellite Radio Web site. www.siriusradio.com.

Space Foundation Hall of Fame Web site. www.spacefoundation.org.

"XM Exceeds Expectations, More Than Doubling Subscribers and Ending First Quarter with over 76,000." XM press release, 1 April 2002.

XM Satellite Radio Web site. www.xmradio.com.

Saving Harbor Porpoises

NASA engineers were more worried about retrieving payloads that splashed down into oceans than they were about keeping harbor porpoises from getting tangled up in fishing nets. But porpoises have benefited from the technology the engineers created.

In the 1960s, engineers at the Langley Research Center developed an underwater location device. The challenge was to make a device that would be strong enough to withstand the high impact of plummeting into the water and it would have to be able to send out a multidirectional beaconing signal for hours after impact.

Dukane Corporation and Burnett Electronics of San Diego, California obtained a license from NASA and set about to improve the design and find commercial applications. Dukane's Seacom division now sells six-and-a-half-inch long devices to sink gill fishing nets to prevent dolphins from getting ensnared in them. Sink gill nets are static nets put out by fisherman to catch bottom-dwelling fish, such as cod and flounder, which swim near the shore. Harbor porpoises are often found in water no deeper than 150 meters and follow these schools of fish—a food source—and porpoises that get caught up in the nets die.

The NetMark device, placed about every 100 yards along a sink gill net, emits a high frequency ping about every four seconds for about thirty days. After that, the pinger can be repowered by replacing its four standard AA alkaline batteries. As long as the dolphins in the area have their internal sonar turned on, they hear the pings as warning beacons and stay away from the net. (When dolphins play, they deactivate their internal sonar and therefore wouldn't be warned not to swim into the nets.)

Several studies have proved the technology's effectiveness. In 1994, a large-scale study off the coast of New Hampshire showed a dramatic reduction in the number of harbor porpoises caught and killed in the Gulf of Maine. According to the National Marine Fisheries Service, "25 harbor porpoises were taken in 423 strings with non-active pingers (controls) and two harbor porpoises were taken in 421 strings with active pingers." The experiments were repeated in different areas over the next few years. During the 1997 study period, "180 hauls were observed with active pingers and 220 hauls were controls (silent). All observed harbor porpoise takes were in silent nets: 8 in nets with control (silent) pingers, and 3 in nets without pingers. Thus, there was a statistical difference between the catch rate in nets with pingers and silent nets."

More recently, a United Kingdom study compared the NetMark pinger to two other methods of trying to reduce porpoise entanglements. "Observers spent 160 days at sea, and monitored 418 strings of nets being hauled." About 40 percent of the nets were equipped with pingers. Only one porpoise was found in the pingered nets, while 18 were found in un-pingered nets. We conclude that at the present time, pingers are the only effective tool available for reducing porpoise bycatch," according to an oral presentation made at the Irish Environmental Researchers Colloquium and posted on the Internet.

The Dukane pingers have also been attached to fishing nets to dissuade sharks, fish, and other marine mammals from swimming into them. They've been used on commercial airliners' flight recorders or black boxes, on hazardous cargo so it can be recovered if it's lost, and to mark underwater sites or relics.

Additional Reading

Culik, B. M., S. Koschiniski, and N. Tregenza. "Reactions of Harbor Porpoises *Phocoena phocoena* and herring *Clupea harengus* to Acoustic Alarms." *Marine Ecology Progress Series* 211 (2001): 255–60.

Dukane Web site. www.dukane.com/seacom.

"Harbor Porpoise (*Phocoena phocoena*): Gulf of Maine/Bay of Fundy Stock." Stock Assessment Report, National Marine Fisheries Service, Washington, DC. December 1999. http://www.nmfs.noaa.gov/prot_res/species/Cetaceans/harborporpoise.html. Accessed 23 April 2003.

Rogan, E., S. Northridge, N. Tregenza, and P. Hammond. "Approaches to Minimising Porpoise Entanglement in the Celtic Sea Gillnet Fishery." Presented at the 12th Irish Environmental Researchers Colloquium, 26 January 2002. http://environ2002.ucc.ie/abstracts/marine13.html. Accessed 8 June 2003.

"Safeguarding Porpoises." *NASA Spinoff 1997.* 1997. http://www.sti.nasa.gov/tto/spinoff1997/er1.html. Accessed 8 June 2003.

Self-Contained Ecosystems

There's no plumbing in space, of course, as well as no trash collection, no supermarkets, and no oxygen resupply. If moons or planets are ever colonized, or if long-term manned space flight becomes a reality, the people living there are going to have to take recycling to a whole new level. Poisonous exhaled nitrogen will have to be eliminated; oxygen will have to be generated. Waste will have to be recycled and food grown.

The intricacies of self-contained ecosystems have fascinated scientists and, thanks to at least one commercial spin-off, they have entertained the Earthbound. EcoSphere Associates, a Tucson, Arizona-based company took technology developed at NASA's Jet Propulsion Lab to create enclosed aquariums. The EcoSphere is a complete self-contained miniature world where the viewer can watch tiny shrimp and algae live in this self-sustaining ecosystem. The shrimp in the EcoSpheres have an average life expectancy of two years, although the maker says it's not uncommon for shrimp populations to thrive for seven years or longer. All that's required is indirect natural or artificial light and the EcoSphere takes care of itself. The shrimp breed, but not to the point of overpopulation because their resources are limited. The shrimp eats the algae, keeping it from taking over the sphere. The shrimps' waste, molted skins and occasional dead shrimp are consumed by tiny bacteria living in the water. So unlike an aquarium, the EcoSphere doesn't need to be cleaned.

EcoSphere Associates sells EcoSpheres in pod or spherical shapes in a range of sizes and the company will even custom design and install larger ecosystems for homes or businesses. In addition to EcoSpheres, the company sells Omnariums and BettaDomes based on the same concept of self-sustaining self-enclosed ecosystems. According to the company Web site, "A unique permeable membrane allows oxygen and carbon dioxide to pass in and out of the Omnarium's shell. This 'breathing' action and minimal feeding helps to create an ideal environment for small freshwater fish and aquatic plants to thrive."

The late astronomer Carl Sagan seemed enthralled with his EcoSphere. He wrote about it in a 1986 *Parade* magazine article, "The World That Came In the Mail." In it, he explained that the similarities between our Earth and these spheres were many but the greatest difference was that humans are able to change our environment, while the shrimp cannot change theirs. "With acid rain, ozone depletion,

The self-contained EcoSphere. Photo courtesy EcoSphere Associates.

chemical pollution, radioactivity, the razing of tropical forests and a dozen other assaults on the environment, we are pushing and pulling our little world in poorly understood directions. Our purportedly advanced civilization may be changing the delicate ecological balance that has tortuously evolved over the 4-billion-year period of life on Earth," he wrote.

While the EcoSphere represents a miniature self-contained world, scientists are also very interested in larger ecosystems. In 1991, people watched in fascination as four men and four women entered the $150 million Biosphere 2 outside of Tucson, Arizona. Biosphere 2 is an enclosed ecosystem built by Space Biospheres Ventures to study how the Earth works. The biospherians, as they were called, lived inside for two years with little interaction and help from the outside world. (Actually, one of the crew left for part of a day for a medical emergency beyond the scope of the physician crewmember.) The original concept was to see if the biological and technical systems built into Biosphere 2 could help with eventual colonizing of the Moon or Mars. But there were problems, including needing to pump additional oxygen into the habitat and the limited ability to grow enough food for the crew. The crew ate a mostly vegetarian diet because milk, eggs, and meat were in short supply. After a second crew stayed inside Biosphere for nearly seven months, the idea of using Biosphere 2 as a human habitat was dropped. It's unlikely it will be a home for people again for the foreseeable future, but it remains an active study ground—and visitors to the site can enter the area where the people lived. While there are no humans living there now, there are more than 3,000 species of living organisms inside—most are insects and plants, and some chickens and fish.

Since 1996, Biosphere 2 has been affiliated with Columbia University and its mission now focuses on research and education. It maintains the original biomes—or different environments. There's a desert, a marsh, a savanna, a rain forest, and an ocean as well as an Intensive Agricultural Biome where the biospherians grew their crops. According to the Web site (www.bio2.edu), the 3.5-acre Biosphere 2 is made of glass, steel and concrete and is 91 feet tall at its highest point. There are 730 sensors inside Biosphere 2 that monitor the air, soil and water every three minutes. Scientists use that information not only to track the health of the ecosystem inside the Biosphere 2, but also to help them understand how the Earth's environment might respond in similar circumstances.

Biosphere 2 is not the only experiment of its kind; Bios-3 is in Siberia. The Russian space program built its first Bios-1 in 1965. Bios-3 was built in 1972 and is completely underground. While it's completely sealed, crew can escape within 20 seconds—a contingency that hasn't yet been needed. The crew area has three sleeping rooms, a kitchen, bathroom, control room, and work area. There were telephones and "viewing ports" to communicate and small airlocks to allow samples to be passed out. The crew's health was monitored constantly. There have been three experiments with crew living in Bios-3; the crews lived there six months, four months, and five months, respectively.

The latest development in this area is being built at the Johnson Space Center in Houston. It's called BIO-Plex (Bioregenerative Planetary Life Support Systems Test Complex). It will be a state-of-the-art eight-chamber test facility.

Additional Reading

Achenbach, Joel. "Biosphere 2: Bogus New World?" *Washington Post*, 8 January 1992, C1.

Ecosphere Associates Web site. www.eco-sphere.com.

Howard, Suzanne. "Out of This World." *Engineering & Technology for a Sustainable World* 7, no. 8 (August 2000): 28.

Johnson Space Center Advanced Life Support Web site. http://advlifesupport.jsc.nasa.gov/.

Maugh II, Thomas H. "8 Pioneers Will Enter Their Own Little World." *Los Angeles Times*, 23 March 1987, 1.

Salisbury, Frank B., Josef I. Gitelson, and Genry M. Lisovsky. "Bios-3: Siberian Experiments in Bioregenerative Life Support." *Bioscience* 47 (October 1997): 9.

Suplee, Curt. "Brave Small World" *Washington Post Magazine*, 21 January 1990, 10.

Tunney, Deborah. "Near Tucson, the Biosphere Lives On" *Washington Post*, 21 March 1999, E4.

Self-Righting Life Raft

When Apollo capsules splashed down in the ocean, astronauts waited in inflatable rafts until helicopters arrived on the scene to pick them up. The problem was that rafts could be overturned by the force of the helicopter's downwash.

Johnson Space Center engineers began working on a method to better stabilize these life rafts, and NASA secured a patent for the heavy ballast design. Around the same time, Jim Givens of Givens Marine Survival Co., Tiverton, Rhode Island, developed a similar system and acquired a license to use NASA's technology.

The result is the Givens Buoy Life Raft, an apple-shaped buoy with a canopy on top and an underwater hemispheric ballast chamber. A one-way flapper valve lets water into the hemisphere chamber, but

Self-righting life raft. Photo courtesy NASA.

no water comes out. The water provides the ballast to keep the center of gravity constant, so the life raft doesn't capsize in rough seas or as the occupants inside move around. The top canopy is bright red, so it can be seen from a distance.

A Givens brochure explains that "water rapidly enters the first stage, 'toroid,' through portholes as inflation chambers force panels apart lending almost immediate stability.... The raft effectively becomes part of and moves with the sea... Because it is water in water, the Givens Buoy Stability System is virtually weightless."

The brochure notes that "if a rogue wave should overturn the raft, the momentum of water in the Givens Buoy Stability System will enable the raft to somersault and reright itself." In fact, company literature states more than 450 lives have been saved due to the life raft. One of the most harrowing tales is recounted on the Web site:

> In August, 1980, four experienced offshore sailors were caught off guard by Hurricane Allen—then the second strongest Atlantic storm on record. Their 30 ton ketch foundered in 170 knot winds with gusts to 190 knots, and the sailors were forced to abandon ship. They inflated their Givens Buoy Raft and climbed aboard. "Once the ballast chamber filled, which didn't seem to take any time at all, the raft settled right down" recalls Bob Harvey, one of the survivors. Ten minutes after entering the raft, the seas calmed and the winds subsided.
>
> Through the canopy porthole, the men could see clear sky at the top of a massive cylinder of clouds. They had entered the eye of the storm! Minutes later the backside of the storm approached and again the men were in the eye wall of the hurricane. They rode 35 foot breaking swells for 42 hours before being rescued by a Norwegian tanker. The men reported that their Givens Raft was at times buried under 6–8 feet of water as the breaking seas and monstrous swells tumbled the raft, but the Givens Buoy Raft always re-righted itself. Harvey recalls "you'd feel it pitch over when we were hit by a wave, but as soon as the pressure came off, it would come back up again. It was like being in a womb. We were floating around in there, sometimes with our feet off the bottom. We didn't feel comfortable, but we did feel secure."

The entire life raft inflates in twelve seconds and the buoyancy chambers fill even faster in rough seas.

The Givens life raft is available in several sizes, ranging from 4- to 12-person capacity. The U.S. Coast Guard uses the rafts on its rescue helicopters, according to the company.

Additional Reading

Givens Life Rafts Web site. www.givenslife rafts.com.

"Self-Righting Life Raft." *NASA Spinoff* (1982): 100.

Ski Boot Improvements

One indelible image of man's first walk on the moon is the photo of *Apollo 11* astronaut Neil Armstrong's footprints on the moon's surface. In a way, Comfort Products, Ltd., of Aspen, Colorado, has been designing and developing footwear with its eye on the moon since the 1970s. NASA technology has led to two major advances in ski boots designed, in part, by Comfort Products.

First, Comfort Products developed rechargeable foot-warming technology. Using the same kind of heating element circuitry that was built into space suits to help regulate the body temperatures of Apollo astronauts, Comfort created these rechargeable devices that many ski boot manufacturers have incorporated into their designs.

More recently, Comfort Products teamed up with ski boot manufacturer Raichle to develop the Raichle Flexon concept in ski boots. The parallels between the demands on astronauts' feet and skiers' feet are striking. Both need to be able to maneuver inside their boots, side to side and front to back, without actually moving their feet. Space researchers had developed an accordion-like extravehicular space suit that allows astronauts to move around more easily. Comfort Products used that same corrugated style for the tongue or front of Raichle's ski and snowboard boots for greater flexibility.

On its Web site, www.raichleusa.com, the company acknowledges the space connection. "The Flexon offers superior maneuverability and more efficient skiing. The technology was originally designed by NASA for use in footwear during the Apollo missions. We adapted it and created a boot that flexes so evenly and smoothly that you'll feel energized on your skis."

Erik O. Giese of Comfort Products, is also the designer of Roller Blades and Easy Spirit shoes. In a story on the technology in the 1995 edition of *NASA Spinoff*, Giese said, "The idea came from the joints in the space suit, which required articulation without distortion of a pressure vessel. It is also similar to a vacuum cleaner hose flexing without destruction, versus a paper towel tube that will bend and crimp when flexed."

The Raichle line of boots uses what it calls a ceramic fit system. A bladder within the boot is filled with hollow ceramic balls that mold snugly around the skier's foot. When the boot is tightened, excess air is squeezed out.

Space suits have undergone a series of changes over the years, according to a NASA Fact Sheet titled "History of Wardrobes for Space." The early Mercury spacesuits were modified jet aircraft pressure suits. They were worn unpressurized because in their pressurized form, movement was very limited. The suits had some give at the elbows and knees, but not much. When the suit was bent at those joints, the pressure around the joints actually increased. Imagine squeezing a balloon in the center. The balloon bulges on either side of your hand.

During the era of the two-man Gemini flights, suits were made more flexible. "Instead of the fabric-type joints used in the Mercury suit, the Gemini spacesuit had a combination of a pressure bladder and a link-net restraint layer that made the whole suit flexible when pressurized," the fact sheet states. The result was improved arm and shoulder mobility.

By the time the Apollo missions were heading for the moon, astronauts needed

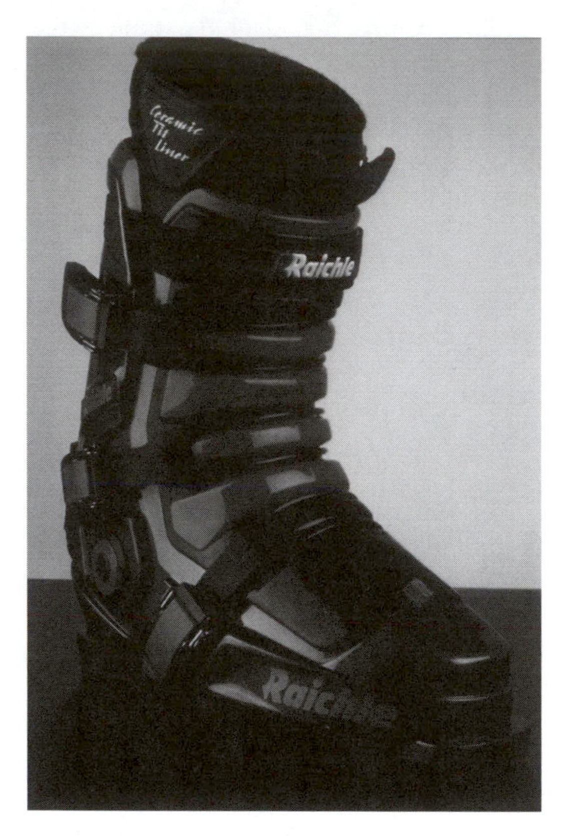

Ski boots. Photo courtesy NASA.

a much more flexible suit. On the surface of the moon, astronauts would need to bend at the knees and waist to be able to scoop up samples and they had to be able to get onto the lunar rover vehicle—the moon buggy. And as astronauts ventured further away from the spacecraft, they needed portable life-support systems incorporated into the suit.

Additional Reading

Raichle USA Web site. www.raichleusa.com.

"Ski Boots." *NASA Spinoff 1995* (1995): 84. http://vesuvius.jsc.nasa.gov/er/seh/pg84s95.html. Accessed 4 June 2003.

"Wardrobe for Space." NASA Fact Sheet, Lyndon B. Johnson Space Center. http://www.jsc.nasa.gov/pao/factsheets/wardrobe.html. Accessed 4 June 2003.

Smart Surgical Probe

A NASA-developed probe could reduce much of the anxiety of waiting for a breast cancer diagnosis. The BioLuminate Smart Probe is expected to be approved and available in late 2003. BioLuminate Inc. acquired the license to produce and market NASA's innovative diagnostic device. The probe was developed at Ames Research Center.

It is a less than one millimeter needle with multiple sensors. The tool could help physicians make a real-time diagnosis without surgery. That would dramatically reduce the number of breast biopsies that are done each year. Currently, that number is nearly 19,000 per week. Only about 15 percent of those biopsies result in a positive cancer diagnosis. As a result, more than 16,000 women per week undergo the procedure needlessly—and at significant emotional trauma and discomfort. In addition, some breast cancer cases go undetected, even with current screening.

According to the BioLuminate Web site, "The difficulty for the physician today, is that the initial screening procedures do not provide any specific information regarding known cancer indicators. A light spot on a mammogram X-ray could be many things in addition to cancer. BioLuminate would like to change that situation by providing six specific measurements of known cancer indicators, so that the physician can make a more informed decision on how to proceed with the patient. This procedure would be utilized only after the initial screening revealed a region of concern. The measurements would be taken simultaneously in real time, through a small 20 to 21 gauge disposable needle that is connected to a computer, as the needle is inserted into the suspicious lesion."

BioLuminate is working with Lawrence Livermore National Laboratory to miniaturize the NASA sensor technology.

Needle biopsies have traditionally not been as sensitive or specific as a surgical biopsy. But with smart software, the SmartProbe is expected to be accurate about 98 percent of the time—the same as a surgical biopsy.

If a physician or woman finds a suspicious lump in the breast, the doctor has to decide whether to send the patient on for a biopsy. The SmartProbe should help provide additional data that could screen out those women whose biopsies will come back negative for cancer. Because the probe has numerous sensors—and since the data is transmitted to a computer right away, the results are displayed in real time. There's no waiting for lab results, so the patient experiences far less anxiety. If the additional data indicates cancer, the physician can immediately refer the woman to treatment. Waiting for pathology lab results can take up to two months.

Specifically, the problem can measure oxygen partial pressure, electrical impedance, temperature, light scattering and absorption properties, deoxygenation hemoglobin, vascularization, and tissue density. These six measurements are known cancer indicators. The physician can measure the tissue prior to reaching the suspicious area and again in the suspicious area. The patient's own tissue measurements serve to set one of the benchmarks for determining whether cancer is indicated.

A future application of the SmartProbe could be delivering drugs, laser heat, or radioactive seeds to the site. It can also be used to monitor the therapies. It could also be used to diagnose prostrate, lung, colon, cervical, and brain cancers.

The company expects that the BioLuminate procedure will cost about $525 per test, compared to $2,620 average for a surgical biopsy.

Neurosurgeons have also looked at the device as a way of better diagnosing brain tumors than the current methods, which can occasionally lead to unintentional bleeding.

Additional Reading

BioLuminate Inc. Web site. www.bioluminate.com.

"NASA Inspired Probe Improves Safety and Efficiency of Brain Surgery." Congress of Neurological Surgeons press release, Park Ridge, Ill., 20 October 1999.

Phillips, Mahoney Diana. "Of Spaceships and Surgery." *Computer Graphics World* (June 1999): 48.

"Smart Surgical Probe Follow-on to Fight Breast Cancer." NASA press release, 14 November 2000.

Rae-Dupree, Janet. "A Quick-Study Cancer Probe" *U.S. News & World Report*, 5 February 2001, 58.

Space Pen

Have you ever tried using a ballpoint pen to write upside down? It doesn't work for very long because most pens rely on gravity. During early space missions, astronauts had to write with pencil because in the microgravity environment, pens wouldn't work.

But that changed during the October 1968 *Apollo 7* flight. Astronauts began using the Fisher AG-7 Space Pen developed three years before. The AG in the pen's name stands for anti-gravity. While the space pen wasn't designed by NASA technology, it was developed with the astronauts' needs in mind and did undergo two years of rigorous testing by the agency before being brought onto a space mission.

Paul C. Fisher developed a pen that used semisolid ink that becomes liquefied only after contact with a rolling ball inside. The ink has the consistency of chewing gum and only flows when the pen is being used. That development solved the problem of leaky ink pens. To make it work without benefit of gravity, Fisher pressurized the cartridge with nitrogen. There's nearly 50 pounds of pressure per square inch, so ink is continuously fed to the ball. As long as the ball is moving when the pen itself is being used, ink will flow. It works at any angle, upside down and underwater, on glossy photos and on greasy paper.

The entire cartridge is enclosed, so there's no evaporation, no leaking, and no wasted ink. The estimated shelf-life of a Fisher Space Pen is 100 years.

Fisher Space Pens have been used on all NASA's manned space missions from *Apollo 7* onward, on the Russian *Soyuz* and *Mir* space flights, and on the ARIANE French space program missions. Having received the boost in publicity from these space flights, the pen was in high demand on Earth. The pen works in all temperature extremes—from –50 to 400 degrees Fahrenheit, so it's been used by the U.S. Air Force, law enforcement agencies, undersea explorers, mountain climbers, and ski and snowboard teams. The pen even had a cameo appearance on the long-running comedy *Seinfeld*.

It is one of the limited number of commercial products that carries the Space Foundation's seal as a certified product. The Space Foundation was founded in 1983 to promote a better understanding of the practical and theoretical utilization of space.

Additional Reading

"Cosmonauts on Mir Become QVC Pitchmen." *Los Angeles Times,* 8 February 1998, A8.

Fisher Space Pen Web site. www.spacepen.com.

Schafer, Sarah. "A Space Pen Odyssey: Pulling Fisher Space Pens Out of a Time Warp Took Some Arm Twisting and Deft Wiring." *Inc.*, 18 June 1996, 74.

Space Foundation Web site. www.spacefoundation.org.

"Space Rose" Perfume

The cosmetics industry learns a great deal from space research. One of the latest developments is a perfume based on a "space rose" note. A note is the word the perfume industry uses to identify the different scents used in a fragrance.

An experiment conducted by astronaut and former U.S. Senator John Glenn on the 1998 Discovery mission was to determine whether microgravity space conditions affected the essential oils of plants. It had long been known that microgravity conditions affect human physiology, and it was assumed there would be some effect on plant physiology as well. International Flavors and Fragrances Inc. (IFF) helped fund the space research to determine what happens to essential oils that provide flavor and fragrance to plants.

IFF's miniature rosebush, Overnight Scentsation, was selected for space travel. It was grown in a special plant-growth research chamber developed by the NASA-sponsored Wisconsin Center for Space Automation and Robotics (WCSAR). Astronaut Glenn collected measurements of chemical compounds found in the roses by inserting a special fiber needle into the blooms. The perfume company was fairly collecting measurements of chemical compounds found in miniature roses by inserting a special fiber needle into the flowers' blooms. What was discovered, to the surprise of the researchers, was that the resulting fragrance was dramatically different—an entirely new scent. In a Marshall Space Flight Center press release, IFF's vice president, Braja Mookherjee, said, "This transformation has created a completely new fragrance that is not of this Earth."

Back on Earth, this new scent was quickly incorporated into perfume. Perfume maker Shiseido has been working on aromachology—the effect that fragrances have on the mind and body—since 1984. In 1964 they developed the original Zen perfume to foster a sense of well-being and peace of mind. That is still being marketed as classic Zen.

But in collaboration with IFF, the company relaunched Zen with the resulting space rose note. The top note, a very light fragrance that lasts for just five to ten minutes, is made up of scents from gentian and water hyacinth. The scent creates "an experience of transparency and freshness," according to the Marshall press release. The middle note, the fragrance that becomes noticeable about fifteen minutes after applying, and generally last for an hour or more, makes use of bamboo, hair cup moss, violet, iris, and the space rose. It invokes silence and nature. The bottom note contains the heavier ingredients and lasts the longest, usually several hours. It uses kyara wood, musk, and oriental amber. The overall experience of the perfume is supposed to be like walking through a Zen garden. Beyond its use in perfume, the oil may also be used as a flavor enhancer in foods.

Space flight has also impacted the beauty industry in other ways. Zero-gravity conditions result in wrinkle-free skin—since there's no gravity pulling fluid in the body downward, therefore the face looks fuller. Cosmetics makers have worked to counteract some of gravity's effects in skin cream products. In an article in *Harper's Bazaar* on NASA technologies' impact on the beauty industry, Loretta Miraglia of Max Huber Research Laboratories traces the relationship back to a rocket-fuel explosion. In 1953, a scientist who was burned there developed a seaweed-based cream that eventually became the company's Creme de la Mer, marketed today by Estee Lauder. "'And today we raid NASA as much as possible,' says Miraglia. La Mer's new Skin-Color Loose Powder contains an ultralight NASA silica developed to insulate rocketships. 'It's easier to create the lightest powder on Earth if you start with the lightest material on Earth,' says Miraglia," in the article.

Additional Reading

Booth, Cathy. "Face-Lift in a Jar?" *Time,* 14 August 2000, 48.

Diamond, Kerry. "Now and Zen from Shiseido." *WWD*, 30 June 2000, 7.

Dougherty, Emily. "Space Case: Space Experiments Inspire the Beauty Industry" *Harper's Bazaar* February 2001, 140.

"Glenn Experiments with Space Rose for Development of New Flavors and Fragrances." NASA press release, Marshall Space Flight Center, 4 November 1998.

"Heavenly Note Helps Center a Down to Earth Perfume." NASA press release, Marshall Space Flight Center, 23 October 2000.

"IFF's Space Research" *Soap & Cosmetics* 75, no. 9 (September 2000): 81.

"IFF's Space Research Results in Unique Fragrance Note." IFF press release, 31 August 2000.

International Flavor and Fragrances Web site. www.iff.com.

Stadium Roofs

When it came time for astronauts to go to the moon, it was clear they'd need a different wardrobe. The spacesuits used for Mercury and Gemini missions were not designed to protect astronauts from the heat of the lunar surface or the jagged rocks found there. It needed to be more flexible than earlier suits were as well.

When Mercury astronauts went into space in their one-man spacecrafts, their suits were essentially high altitude pilot's pressure suits. The idea was that these suits could be pressurized if there were ever a loss of pressure in the spacecraft cabin. For the Gemini missions, space suits underwent some changes. With two people in a tiny spacecraft, it was now even more important that the suits allowed greater flexibility. The Gemini suits were made from multiple layers and arm and shoulder mobility was improved. The first space walk happened during *Gemini IV*'s flight in June 1965. Astronaut Edward White's life support system was connected to the capsule's system by an umbilical cord.

While the earliest space walks showed that astronauts could survive outside of the relative safety of the spacecraft, walking on the moon would require reoutfitting astronauts. Astronauts would have to be able to carry portable life support systems and the suits would have to be more flexible still. Astronauts stepping onto the moon's surface would need to stoop and bend to pick up rocks or other samples from the moon's surface, do their experiments, and ride in the lunar rover. The fabric for the outer layer of the space suit would have to be noncombustible, strong, durable and lightweight.

In the 1970s, Owens-Corning was developing a glass-fiber yarn that could be woven into a fabric. The fabric was then coated with Teflon, which added strength and durability. The fabric also repelled moisture. It was used in spacesuits for all Apollo missions.

It wasn't long before that same fabric found a commercial use—as roofing material for sports stadiums, shopping malls, and airport terminals. "Pound for pound, the material is stronger than steel and weighs less than five ounces per square foot," according to an article in a *NASA Spinoff*. These factors combine to lower initial costs and speed construction.

Birdair Inc. of Amherst, New York, markets the material. On its Web site, Birdair shows how the material can be used in custom tensioned structures. In these structures, the roofing material is supported by a network of cables or pylons. These range from smaller entrance canopies to domed stadiums, retail centers, amphitheaters and transportation depots. This construction system can be used with retractable roofs, an advantage for sports facilities where the roof can be pulled back on nice days. Having a retractable roof also means that grass can grow in the dome.

Birdair states that three-quarters of the sports domes in North America have been constructed by the company. Denver International Airport and Chicago's Navy Pier are among other projects featuring Birdair's fabric. The Denver airport project is cited as an example of how well the roofing material holds up under heavy snows and strong winds. The design is not indestructible, though. A March 2003 storm dumped 30 inches of snow on the Denver airport roof and the weight of the snow put a 40-foot tear in the material.

Beyond its strength, the material has other advantages, states the company. For example, in retail settings "Natural light diffused by the membrane reduces glare, hot spots and merchandise fading, and keeps energy costs low." River Falls Mall in Clarksville, Indiana, and Crystal River Mall in Crystal River, Florida, are two malls that feature the roofing material.

Air supported structures are essentially fabric membrane envelopes that are supported by pressurized and heated air. The advantages are a column-free interior space, use of natural daylight, and a cost savings of about one-third over conventional construction materials. The Tennessee Titans practice facility is one example, as are numerous pool enclosures, tennis enclosures, golf domes, and fitness centers.

Additional Reading

Birdair Web site. www.birdair.com.

"Fabric Structures." *NASA Spinoffs: 30 Year Commemorative Edition* (1992): 99.

Toy Gliders

How many people does it take to make a foam airplane that can fly well?—as it turns out, a lot. Hasbro, Inc., the Pawtucket, Rhode Island, toy company that makes Nerf products, developed a foam glider. To make it so that kids could fly it a decent distance and make it do loops, the company contacted NASA.

As a Hasbro design director said in an article in the 1997 issue of *NASA Spinoff,* "Who knows better how to make things fly than NASA?" Regional technology transfer center staff directed the company to the Langley Research Center with its wind tunnel. The wind tunnel is used to test aerodynamics of airplanes and cars.

They enlisted the help of Ray Whipple, a Langley wind tunnel manager who specializes in testing models of military fighter planes. According to a story in the *Newport News Daily Press,* "When Whipple first heard about the project, he thought of the Nerf Ball, and said to himself, 'They'll never get any of that to fly.' "

The group helping Hasbro called in two NASA Langley retirees who also happened to be model airplane specialists. Together, the group helped Hasbro designers figure out such details as the best angles for tail fins and where to locate the wings on the glider.

The group spent a day together. What Hasbro wanted was a foam glider that had no moving parts—so kids wouldn't have to make adjustments to wings or tail fins or anything.

The *Newport News Daily Press* detailed the meeting that toy designers and the NASA team had. The Hasbro team had sent sheets of their Nerf foam and that afternoon, they "broke out the foam and the glue and joined the NASA team in piecing together a new set of prototypes. . . . The group tested their new designs, first in a hallway, then out on the lawn in front of Langley's library. The sight prompted passersby to stop, stare, and, in some cases, walk over for a closer look."

What came out of the meeting were several versions of the toy gliders, including the Super Soaring Glider—a long-distance, high-performance flyer and the Ultra Stunt Glider, which does acrobatic tricks such as loops and turns.

Additional Reading

"Hasbro Launches New Toy with NASA's Help" *Innovation* 4, no. 3 (July/August 1996).

Sterling, Richard. "Toy Story: NASA Meets Nerf." *The Newport Daily News* (Newport, R.I.), 29 February 1996.

"Toy Gliders." *NASA Spinoff 1997* (1997): 75.

Toy gliders. Photo courtesy NASA.

Traffic Intersection Safety

In 1978, Jim Davidson was nearly hit at an intersection by a fire truck on its way to an emergency call. Davidson hadn't seen the truck coming. Rattled, but uninjured, he was one of the lucky ones. Firefighters, paramedics, and police officers responding to emergency calls are involved in thousands of accidents each year. In 1997, there were 15,000 of these accidents, resulting in 8,000 injuries, 500 deaths and millions of dollars worth of property damage, according to the *Federal Highway Administration Newsletter*. Simply getting to the scene of an accident or emergency has risks for first responders. Of the firefighters who die in the line of duty, about 40 percent of them are killed on the way to the emergency. According to the U.S. National Highway Traffic Safety Administration, in 1999 there were 73 percent more deaths during emergency calls, when lights and sirens were operating, than when the trucks were driven at other nonemergency times.

His own near-miss collision got Davidson to thinking about ways to provide better alerts to drivers that emergency vehicles are coming. Ultimately, Davidson created the Emergency Vehicle Early Warning Safety System or E-ViEWS, a high-tech visual warning and communications system. The system was developed with the help of the Technology Affiliates Program at NASA's Jet Propulsion Laboratory (JPL) in Pasadena, California. A DOT newsletter detailing the first trial of the system, explains "Under JPL's Technology Affiliates Program, large and small businesses are able to tap the specialized expertise of JPL engineers to solve particular technical challenges. As James Rooney, the Technology Affiliates Program director notes, 'A very important part of the NASA/JPL mission is not only to explore the universe and to develop the technology that meets that mission but also to see if we can work with companies to find good uses of that technology, especially when it includes humanitarian and global impacts.' "

A prototype of the system was created in just a few months. Here's how it works. An emergency vehicle has a satellite-linked transponder in it. As the driver flips on the siren, the transponder sends a microwave message to change traffic lights as it approaches them. The signal is strong enough to work on traffic lights up to a kilometer away. To announce the arrival of an emergency vehicle, the system has another step. Large, four-square-foot illuminated signs are mounted at these intersections that feature animated icons of the intersection. When the traffic lights change to red, the signs display from which direction the emergency vehicle is coming. The lighted emergency vehicle icon appears as if it's moving toward the intersection. After the emergency vehicle has passed, the traffic lights resume their normal pattern. Check out the E-ViEWS Web site (http://eviewsinc.com) to see what drivers see.

The warning system actually kicks in as soon as an emergency vehicle is dispatched. The intersection interface holds the information in its memory and waits to be contacted by a transponder in the emergency vehicle.

The system was tested in Monrovia, California, a city just seven miles away from JPL. E-ViEWS provided about $600,000 worth of equipment if the city's emergency response crews agreed to test the system for three months. JPL oversaw the testing as well. In the spring of 2001, ten intersections were wired and twenty police cars and ten fire emergency vehicles were outfitted.

In the JPL press release announcing the system test, Monrovia's police chief Joe Santoro said, "When responding to emergencies with red lights and sirens, emergency vehicles present a serious traffic hazard to themselves and other vehicles and pedestrians while passing against traffic through an intersection. Confusion, inattention, mobile phones, car radios, hearing impairment, distracting children and failure to hear sirens and see flashing lights are just a few of the many causes of serious accidents that result in multi-million-dollar lawsuits against cities and states."

The system was put to its first serious test in July 2001. According to the *Federal Highway Administration Newsletter*, emergency crews received a call that a resident had been attacked by a vicious dog and was bleeding seriously. "By using all E-ViEW-equipped intersections, rescuers were able to shave several minutes off their response time, minutes that the fire department later reported had made the difference between the life and death of the victim." A system engineer said, "The wounds inflicted were life threatening and every minute counted. I feel the E-ViEWS Safety System contributed greatly in the saving of a life."

E-ViEWS Safety Systems, formed in October 1998, is also working on similar warning systems for railroad crossings and other traffic situations. The company's Web site mentions the technological leaps in traffic control that have occurred in recent decades. "Twenty-five years ago, few signalized intersections had walk/don't walk signs. Local governments recognized the improvement in safety, better traffic flow and reduced accidents through the installation of the device."

Additional Reading

"Emergency Vehicle Early Warning System to Be Tested in Monrovia." NASA press release, Jet Propulsion Laboratory, 13 December 2000.

E-ViEWS, Inc. Web site. http://eviewsinc.com.

"Green Lights All the Way Could Save Lives." *CNN.com*. 20 May 2001. http://www.cnn.com/2001/US/05/20/highspeed.chases/index.html. Accessed 5 May 2002.

"Monrovia, Calif., Uses Space Program Technology to Make Street Intersections Safer during Emergency Calls." *Business Wire* (25 April 2001): 0332.

"Monrovia, California: Emergency Vehicle Preemption and Visual Warning System" *Federal Highway Administration Newsletter*, 29 May 2002. http://www.eviewsinc.com/index.cfm?page=projects&i4=92. Accessed 7 June 2003.

"Public/Private Partnership Will Make Monrovia the First City in the Nation to Install an Emergency Vehicle Intersection Early Warning." Monrovia Police Department press release, 12 December 2000.

"Traffic Lights That Think Ahead." *Maclean's*, 8 January 2001, 44.

Video Camera Improvements

Much of space research involves simply looking at distant objects. Solar flares on the sun, storms on planets and the flight of comets are beyond reach except through a telescope. So the quality of the image produced by the telescope is critical for scientists.

The trouble is, Earth-bound telescopes—even the most sophisticated ones—are subject to the slightest tremble. Though undetectable to the eye, tiny movements in a telescope can ruin a video of an object millions of miles away. That was a problem for scientists who use a telescope to shoot video images because a shaky video is fuzzy, blurry, and difficult to decipher. It also interfered with the work of meteorologists studying the movement of weather systems from satellite video images.

Two scientists at the Marshall Space Flight Center in Huntsville, Alabama, though, discovered a way to stabilize the images they recorded of the sun and of weather systems on Earth. Dr. David Hathaway, a solar astronomer and Paul Meyer, a meteorologist and computer scientist, used a computer algorithm that first corrects problems in each frame of the video, and then combines similar frames to create a single, clearer image.

Their discovery stayed inside NASA until 1996, when it was used in an unusual circumstance. The Federal Bureau of Investigation came to the Marshall Center after the bombing in Atlanta's Centennial Park during the Summer Olympic Games. The FBI had amateur videos taken at Centennial Park, and asked if NASA could improve the images in hopes it might find a suspect.

In an interview with ABC News, Hathaway explained what he and Meyer were able to do: "They'd have several seconds of video, so we would go through and add a dozen frames together, then move over and add a dozen frames together," he said. "By adding them together, you could start seeing this person walk around, which was not at all apparent in the original video."

The two scientists worked on a number of cases with law enforcement after the Atlanta investigation. In one involving a kidnapping and murder of a Minnesota teen-ager, Hathaway helped improve video from a security camera. The tape was later used as evidence in the trial of a man convicted of the murder.

VISAR software used on a single video frame taken at night. Photo courtesy NASA.

Working with the FBI helped the two scientists refine their technique into a product they call VISAR, for Video Image Stabilization and Registration. When it was ready to license, security companies jumped at the chance to use the technology. Intergraph Government Solutions of Huntsville uses VISAR in a Windows-based program that it markets to police departments for use in video surveillance, sting operations, and on dash-mounted video cameras. The Intergraph system has a variety of tricks—it can take a video shot in dim light and brighten it, for example, zoom in, or follow a single subject no matter where it moves in the frame.

A second company, BarcoView America, of Duluth, Georgia, has licensed the technology as well. The company makes video imaging devices used in areas such as medical imaging and air-traffic control.

This technology developed for studying large objects has been used to study cells as well. The Casey Eye Institute at the Oregon Health Sciences University in Portland, Oregon, used it to improve video of cell movements in the eye. "After NASA enhanced the video, we could see cell movements inside the eye that were undetectable before," said the institute's Dr. Stephen R. Planck in an interview with NASA's *Innovation* magazine.

Hathaway and Meyer told *Innovation* they hope to see the VISAR system used in home-video cameras. If used in real time it could automatically correct the video image when the user tilts or jiggles the camera, and improve single-frame images. "It's amazing to me that software we invented has the potential to be used everyday in home computers across America," said Meyer.

Additional Reading

Fox, Barry. "NASA Goes to the Rescue of Shaky Home Videos." *New Scientist* 162, no. 2187, (22 May 1999): 6.

"Innovative Solution to Video Enhancement." *NASA Spinoff 2001* (2001): 103.

"The New Standard in Forensic Video Analysis." Intergraph Government Solutions Web site. http://www.intergraph.com/govt/hardware/vas/va.asp. Accessed 13 May 2002.

Space Foundation Web site, 2001 Hall of Fame inductees. http://database.spacefoundation.org/hof/hall_of_fame.cfm.

"Video Clarification Better Than Ever." *NASA Aerospace Technology Innovation,* July/August 1999, 9–11.

Virtual Reality

Virtual reality has become a tangible reality, thanks to NASA-developed research on telerobotic control of automated systems in space. Virtual reality, still an emerging technology, allows computer users to immerse themselves into a realistic-looking computer-generated environment. These 3D graphic environments can be explored and manipulated by the user.

Virtual reality has been evolving at NASA's Ames Research Center since 1984. The first Virtual Interface Environment Workstation was a head-mounted stereoscopic display that allowed the operator to step into a scene and interact with it.

Many people are familiar with virtual reality as a really cool computer game. But it has tremendous applications in planetary explorations, medicine, flight, and airport safety. NASA's Mobile Aeronautics Education Lab (MAEL) has a Virtual Reality Station in which students can "fly" various aircraft and simulate NASA research studies in areas such as remote sensing and Earth observation.

To make virtual reality truly interactive, the user is immersed into the virtual world. This can be done through a headset that shows high-resolution stereo images. Virtual reality also brings true-to-life sounds to the user, for example, the sound of rushing wind in a flight simulator or the scream of an engine in race car simulation. The user interacts with the virtual environment in some way. VPL Research in Redwood City, California, developed a glove lined with sensors that communicates with a computer for Ames. Called the DataGlove, the sensors transmit hand movements to the virtual reality environment. The user can pick up a rock on the surface of the moon and feel it.

A November 1997 paper titled "About Virtual Reality and Its Use in the Mobile Aeronautics Education Laboratory," talks about the importance of virtual technology in NASA's overall missions, for example, virtual reality-assisted astronauts making repairs to the Hubble Space Telescope. "Researchers at Johnson Space Center (JSC) created a highly realistic model of the Hubble space telescope and all of the instruments they would install or replace. During the training, the astronauts stepped into a virtual environment created in a JSC laboratory through a Head Mounted Display and a DataGlove to test the procedures and methods they would actually perform while on orbit. Although the researchers could not provide the astronauts with the experience of reduced gravity while in a virtual environment, the visual computer-generated 3-D images which they saw and were able to manipulate were highly realistic and very effective."

Researchers are developing a parallel technology called Scientific Data Visualization, which transforms data into full-color computer animation. In other words, scientists can visualize how air molecules move around an aircraft. Indeed, researchers at the Ames Research Center have developed a virtual wind tunnel to do just that while researchers at the NASA Jet Propulsion Lab developed the ability to visualize the surface of a planet much the same way.

In June 2000, NASA's two-story airport simulator called "FutureFlight Central" won an award for the most significant contribution to aviation safety. The simulator allows planners to create an airport so that planners and controllers can test new designs and modifications of existing airports. It's a full-scale, 360-degree simulator in which planners can test runways, landings, ground traffic, and other airport factors.

In a press release announcing the award, Dr. Paul Kutler, deputy director of the NASA Ames Information Systems Directorate, said, "NASA's FutureFlight Central hopes to save airports costly design errors by permitting planners to easily experience different, highly realistic versions of their airport designs and, most importantly, observe how real people work inside these future environments." In the simulator, planners can make changes to the airport design almost instantaneously.

Virtual reality has also contributed to medical training. Medical students and physicians can "practice" new procedures in a virtual reality setting. In another recent development, virtual reality is being used as a kind of physical therapy program. Parkinson's disease or stroke patients wear a small computer with an optical display that clips to eyeglasses. The device was created by a computer science professor Yoram Baram, a former NASA engineer, who learned that Parkinson's patients walked better on tiled floors. By providing virtual targets to help patients stabilize themselves, the device can help improve the patients' walking speed and steadiness.

Baram, now at Technion-Israel Institute of Technology, explained in an article on the institute's computer sciences page: "The image reacts to the patient's motions just like in real life...For example, when the patient stands in place, the virtual floor doesn't move, but when he begins to walk, the floor starts moving beneath him. When he turns, the image of the floor also turns. Yet all the while the patient feels like he is walking on a steady floor."

The article continues, "The idea for the project was sparked 12 years ago while Baram was designing a mechanism for NASA to navigate low-flying helicopters around obstacles such as trees, buildings and electrical poles. The concept of the design, which Baram later applied to the medical device, is that objects appear to expand as you approach them. 'A person who is walking uses visual images to navigate himself so he doesn't collide...While a healthy person has internal mechanisms to help balance himself, Parkinson's patients are deficient in this area. But they can be helped by visual cues like tiles that operate through biofeedback.' "

As graphics and robotics improve to allow greater realism, virtual reality will become even more important.

Additional Reading

Feiner, Steven K. "Augmented Reality: A New Way of Seeing; Computer Scientists Are Developing Systems That Can Enhance and Enrich a User's View of the World." *Scientific American*, April 2002, 48.

FutureFlight Central Web site. http://ffc.arc.nasa.gov.

Garbi, Jill. "Virtual Reality for Aiding People with Movement Disorders." http://www.cs.technion.ac.il/~baram/tiles.html. Accessed 7 June 2003.

"NASA Virtual Reality Airport Simulator Wins Air Safety Award." NASA press release, 29 June 2000.
Phinisee, Tamarind. "Researchers Sink Teeth into Virtual Reality Program." *San Antonio Business Journal*, 31 August 2001, 25.
"Virtual Reality" *NASA Spinoffs: 30 Year Commemorative Edition* (1992): 90–91.

Weather, Climate, and Geologic Mapping

Satellites have made a world of difference to meteorologists and scientists alike. When the first weather satellite was launched in the early 1960s, it immediately became apparent that the data provided about global weather phenomena was like nothing on Earth. Suddenly it was possible to see weather patterns developing, making three- to five-day weather forecasts possible.

In the following decades, NASA launched a variety of satellites that were capable of seeing new features. Landsat became the world's first civilian land-imaging satellite sending back information about land surface features and vegetation. The Earth Radiation Budget Experiment in the 1980s studied how the Earth absorbed and reflected solar radiation, and during that time, NASA's Total Ozone Mapping Spectrometer sent back information about the hole in the ozone layer over the Antarctic. It was NASA's Upper Atmosphere Research Satellite that confirmed that industrial chemicals were the source of ozone-destroying compounds.

Today, satellites measure weather and map clouds, land surface, winds at ocean surface, stratospheric gases and aerosols, ocean surface topography, the forest canopy, and ice sheet topography. Other airborne measurement devices also contribute to the knowledge base. For example, scientists at the Jet Propulsion Lab were able to map topography and changes in topography from instruments aboard an aircraft just after a volcano erupted. The resulting information will help scientists predict where different types of lava will flow. Thermal infrared tracking can be used to map ground temperatures or to map the amount of sulfur dioxide in volcanic plumes. Changes in either of these areas could signal impending volcanic activity.

According to NASA's Earth Sciences Enterprise, major new developments are coming. In 25 years, satellites and spacecraft will send back the information, and supercomputers will process it well enough to develop 10-year climate forecasts, 15- to 20-month El Niño predictions, 12-month regional rain rates, 60-day volcano warnings, 10- to 14-day weather forecasts, 7-day air quality notifications, 5-day hurricane track predictions to within 30 kilometers, a 30-minute tornado warning, and possibly advanced earthquake forecasting.

Of course, it's not just a matter of seeing what's happening, but also understanding it and using that information to forecast. And that takes considerable computer capabilities. A May 2, 2002, Science@NASA article on climate modeling notes that NASA computer engineers at Ames Research Center are producing a ten-fold improvement in computing power. For example, a supercomputer at Goddard Space Flight Center has a peak performance of 409 gigaflops—a gigaflop is a billion calculations per second. The Ames supercomputer will perform even faster.

The story explains that as a result of a four-year partnership with computer maker Silicon Graphics, Inc., the scientists developed two new technologies. One is called single-image shared memory. "In this design, all of the supercomputer's memory is used as one continuous memory space by all of the processors. (Other architectures distribute the memory among the processors.) This lets the processors exchange the messages needed to coordinate their efforts by accessing this 'common ground' of memory. This scheme is more efficient than passing the messages directly between the processors, as most parallel supercomputers do."

The second technology is called multilevel parallelism. The article explains, "Software made using this tool can use the common pool of memory to break the problem being solved into both coarse-grained and fine-grained pieces, as needed, and compute these pieces in parallel. The single memory space gives more flexibility in dividing up the problem than other designs in which the memory is physically divided among the processors."

Here's how the developments improve climate modeling. The atmosphere and oceans are divided into a 3-D grid. Each box within the grid is assigned values for temperature, moisture content, chemical content, and other measurements. "...[T]hen the interactions between the boxes are calculated using equations from physics and chemistry. The result is an approximation of the real system. With more computing power available, more of the physics of the real climate system can be incorporated into the models, and the atmosphere can be divided into more, smaller boxes. This makes the models more realistic, and the predictions they will produce will be of more interest on a regional scale."

As a result of developments like this, climate modeling will become a more precise science.

Additional Reading

Barry, Patrick L. "Modeling Climate at Warp Speed." *Science@NASA,* 2 May 2002. http://science.nasa.gov/headlines/y2002/02may%5Fsupermodel.htm. Accessed 10 May 2002.

"Exploring Our Home Planet: Strategic Plan." NASA Earth Science Enterprise, November 2000. http:// www.earth.nasa.gov .Accessed 10 May 2002.

Wildfire Control

When it comes to fighting forest fires, an eye in the sky can be very helpful. NASA technology has made pinpointing fires and relaying pertinent information to firefighters on the ground easier through with several approaches.

NASA's Jet Propulsion Laboratory (JPL) has been working on systems to help U.S. Forest Service firefighters better detect and track fires using infrared fire detection technologies. The systems have become lighter-weight, faster, more sensitive, and able to scan more ground. In its 1990 *Spinoff* magazine, NASA reported on a JPL-developed fire detection system and since then, many developments have taken place.

In the early 1980s, engineers and scientists at JPL devised the Fire Logistics Airborne Mapping Equipment (FLAME) system, an aircraft-based infrared scanning system that was able to pinpoint a fire's location. The equipment was large—it weighed 87 pounds and was about three feet tall by two feet wide. It's still being used, but the data storage trunk is no longer needed, making it a much lighter and smaller system. The system has been used to identify hot spots not visible as flames to help firefighters find areas that are likely to flare up.

Infrared scanning systems are better tools for documenting fires than photographs, since images can be obscured by thick smoke. Plus, infrared's ability to map hot spots identifies potential danger spots.

In 1996, NASA began experimenting with real-time delivery of data to firefighters using Internet technology. That's when NASA used an ER-2 aircraft based at Ames Research Center in Mountain View, California, to fly over wildfires and send back real-time images. The plane became the first to use the Satellite Telemetry and Return Link (STARLink) system to relay information through the Internet.

Then came Altus II, a specially designed, remote-controlled airplane that flies at 60,000 feet, takes photos of fire scenes and uses the Internet to transmit them nearly instantaneously to firefighters on the ground. Because the plane is uninhabited, it can fly longer than a plane with a crew. In tests, the plane stayed up for as long as 24 hours. It uses a digital multispectral scanner that detects fires and hot spots even through thick smoke. Altogether, there is about 200 pounds of equipment installed on Altus II.

The images sent from Altus II are compared to maps and that information is posted on the Internet about 15 minutes later. Altus II flies higher than other airplanes, so information gathered from this plane covers a much wider area. Between its broader view and longer flying time, the plane provides much more information to firefighters.

In 2001, NASA launched its Terra satellite with a sensor package specifically dedicated to detecting fire outbreaks. It uses the NASA-designed Moderate Resolution Imaging Spectroradiometer (MODIS), which can sense thermal infrared energy given off by wildfires. NASA and University of Maryland scientists developed the software that turns this data into maps. The system should prove to be much more efficient than either weather satellites—which are designed to show clouds—or airplanes, which can cover a much smaller area. It can also go where reconnaissance planes cannot. If smoke is too heavy, it can be too dangerous to send a plane up. MODIS senses colors in the visible and infrared spectrum. It not only sees flames and infrared radiation, it can also sense how hot a fire is.

The satellite beams images of wildfires daily, within hours of passing over the wildfire-prone Western region of the country. That transmission time is being reduced. These images are sent to the Forest Service. Computer users can see the updated maps for themselves at www.nifc.gov/firemaps.html.

And after the terrorist attacks on the World Trade Centers on September 11, 2001, NASA was asked to lend its expertise and equipment. NASA and the U.S. Geological Survey worked together to get a NASA-owned airplane to fly over the rubble to help identify minerals in the air and scan for hot spots. The Airborne Visible Infrared Spectrometer (AVIRIS) is a remote-sensing unit designed to help explore planets. Planes were grounded immediately after the attacks, but special clearance was given for these surveys. A NASA team flew four missions over the area in the two weeks following the attack. Information gathered by these flights and correlated with air and dust samples from around the World Trade Center region pinpointed 34 fires, previously undetected, burning deep in the rubble, and provided valuable information about air quality. The caustic dust cloud contained asbestos, a high pH level, and many heavy metals and minerals.

Additional Reading

"An Eye in the Sky to Help Snuff Fires." *Business Week,* 10 September 2001, 91.

Daukantas, Patricia. "NASA Tests Remote Craft to Gather Wildfire Data." *Government*

Computer News 20, no. 28 (17 September 2001): 17.
"Forest Fire Mapping." *NASA Spinoff 1990* (1990): 106.
Higgins, Amy and Sherri Koucky. "Fighting Fires from Space." *Machine Design* 73, no. 21 (8 November 2001): 41.
"High Tech Fire Fighting" *NASA Explores.* 13 December 2001. http://nasaexplores.com/show2_articlea.php?id=01-091. Accessed 7 June 2003.
"NASA's Information Technologies Aid California Fire Fighters." NASA press release, Ames Research Center, 28 August 1996.
"Satellite Helps Hasten Response to Wildfires." *Engineering & Technology for a Sustainable World* 8, no. 10 (October 2001): 4.
Schneider, Andrew. "Ground Zero's Roiling Dust Cloud Filled U.S. Scientists with Sense of Urgency." *St. Louis Post-Dispatch,* 10 February 2002.

ZeoponiX

Imagine living in the International Space Station for weeks, maybe even months. Among the comforts of home missed the most are the taste of a vine-ripened tomato or a crisp green bean. Plants provide oxygen, food, and can help recycle wastes; therefore, growing plants would be a critical element of long-term space travel. But ordinary potting soil is too heavy to lug into space. So in the late 1970s scientists began working on developing a suitable growing medium.

After years of testing, NASA developed a soil amendment and fertilizer from zeolite, a crushed rock mined from ancient volcanic ash deposits. It's lightweight and more porous than ordinary soil, so it holds moisture. One of its characteristics is a high cation exchange capacity, meaning that plant nutrients such as nitrogen and potassium can be held to the zeolite. These nutrients are released slowly, fertilizing the roots of the plants.

Hydroponics is the science of growing plants in a nutrient-rich solution or moist material other than soil. To describe this new growing medium, NASA combined the words zeolite and hydroponics to create zeoponics. In 1992, NASA's Johnson Space Center contracted Richard D. Andrews of Boulder Innovative Technologies (BIT) to help develop zeoponic science and materials for future space use as well as Earth-bound uses. In space, plants could be grown directly in the nutrient-charged zeolites. But on Earth, the charged zeolites can be mixed with soil or other materials like peat or perlite. BIT spun off a company called ZeoponiX, Inc. and has an exclusive license to commercialize the technology. Andrews is founder of ZeoponiX.

Its first commercially available zeoponic product is called ZeoPro a fertilizer primarily for use in golf courses. In a two-year test at Colorado State University and another test at Cornell University, grass grew much faster and thicker in zeoponic-enriched plots than it did in comparative sand-peat root zone control plots. Neither the kind of grass, nor whether the test plot was started from seed, sod, or sprig seemed to matter. Even when only half the normal fertilizer was applied to zeoponic plots, the root mass development was double the fully fertilized comparison plot.

Those kinds of study results show that ZeoPro is ideally suited for turf that gets rough treatment, such as soccer fields or golf courses. The product has been the subject of numerous stories in golf course management magazines for its ability to revive worn turf. Some golf course workers have taken to calling it space dust.

However, ZeoponiX is hoping that the market is much bigger. They are marketing the material to greenhouses, commercial landscapers, and home gardeners. The company's Web site shows comparison photos of marigolds, geraniums, and impatiens grown with conventional fertilizers and with ZeoPro Super. Similarly, tests done with tomatoes showed that a backyard gardener adding ZeoponiX could get significantly higher tomato yields—up to 92 percent higher, depending on the ZeoponiX mix used—than using other routine fertilizers. In 1997, the company ran a test with a commercial tomato grower. The result was that the plants grown in ZeoponiX mix produced two to three times the number of fruit. And the fruit was heavier than those fertilized in a traditional way.

In three years of testing ZeoPro in rice production in Malaysia, results have shown from 40 percent to more than 100 percent higher rice yields. "This is a very exciting result since rice is one of the most important food crops in the world," said Richard Andrews.

There are nearly fifty types of zeolites with varying physical and chemical properties. Some can help plant growth while others make excellent filtration media. Even the same type of zeolite can have different properties, depending on the environment where it was found. Zeolites have applications far beyond horticulture. They can be used as biofilters in aquaculture. For example, in fish hatcheries, zeolites can remove toxic ammonia from the water. In agriculture, they can be used for odor control, confined animal environment control, and even as livestock feed additives. Similarly, they are used to control household or pet odors. Industrial applications for zeolites include absorbents for oil and chemical spills and gas separations. They are used for hazardous site remediation or decontamination to absorb radioactive waste. They can be used to treat or filter water in swimming pools and to remove toxic heavy metals or ammonia in wastewater treatment facilities.

Additional Reading

Allen, E. A., and R. D. Andrews. "Space Age Soil Mix Uses Centuries-Old Zeolites." *Golf Course Management* 65, no. 5 (May 1997).

Andrews, R. D., A. J. Koski, J. A. Murphy, and A. M. Petrovic. "Zeoponic Materials Allow Rapid Greens Grow-In." *Golf Course Management* 67, no. 2 (February 1999).

Andrews, Richard, James Shaw and Dr. James Murphy, "Zeoponic Turf Root Zone Systems: Bringing NASA-Developed Technology to Sports Turf." *SportsTurf,* April 1999.

Olgeirson, Ian. "'Space Fertilizer' Product Launched for Earthly Use." *Denver Business Journal,* 17 July 1998, A4.

ZeoponiX Web site. www.zeoponix.com.

Appendix A

A Brief History of NASA

by Stephen J. Garber and Roger D. Launius

Launching NASA

"An Act to provide for research into the problems of flight within and outside the Earth's atmosphere, and for other purposes." With this simple preamble, the Congress and the president of the United States created the National Aeronautics and Space Administration (NASA) on October 1, 1958. NASA's birth was directly related to the pressures of national defense. After World War II, the United States and the Soviet Union were engaged in the Cold War, a broad contest over the ideologies and allegiances of the nonaligned nations. During this period, space exploration emerged as a major area of contest and became known as the space race.

During the late 1940s, the Department of Defense pursued research, rocketry, and upper atmospheric sciences as a means of assuring American leadership in technology. A major step forward came when President Dwight D. Eisenhower approved a plan to orbit a scientific satellite as part of the International Geophysical Year (IGY) for the period, July 1, 1957 to December 31, 1958, a cooperative effort to gather scientific data about the Earth. The Soviet Union quickly followed suit, announcing plans to orbit its own satellite.

The Naval Research Laboratory's Project Vanguard was chosen on 9 September 1955 to support the IGY effort, largely because it did not interfere with high-priority ballistic missile development programs. It used the nonmilitary Viking rocket as its basis while an Army proposal to use the Redstone ballistic missile as the launch vehicle waited in the wings. Project Vanguard enjoyed exceptional publicity throughout the second half of 1955, and all of 1956, but the technological demands upon the program were too great and the funding levels too small to ensure success.

A full-scale crisis resulted on October 4, 1957, when the Soviets launched *Sputnik 1*, the world's first artificial satellite as its IGY entry. This had a "Pearl Harbor" effect on American public opinion, creating an illusion of a technological gap and provided the impetus for increased spending for aerospace endeavors, technical and scientific educational programs, and the chartering of new federal agencies to manage air and space research and development.

More immediately, the United States launched its first Earth satellite on January 31, 1958, when *Explorer 1* documented the existence of radiation zones encircling the Earth. Shaped by the Earth's magnetic field, what came to be called the Van Allen Radiation Belt, these zones partially dictate the electrical charges in the atmosphere and the solar radiation that reaches Earth. The U.S. also began a series of scientific missions to the Moon and planets in the later 1950s and early 1960s.

A direct result of the *Sputnik I* crisis, NASA began operations on October 1, 1958, absorbing into itself the earlier National Advisory Committee for Aeronautics, which included 8,000 employees, an annual budget of $100 million, three major research laboratories—Langley Aeronautical Laboratory, Ames Aeronautical Laboratory, and Lewis Flight Propulsion Laboratory—and two smaller test facilities. It quickly incorporated other organizations into the new agency, notably the space science group of the Naval Research Laboratory in Maryland, the Jet Propulsion Laboratory managed by the California Institute of Technology for the Army, and the Army Ballistic Missile Agency in Huntsville, Alabama, where Wernher von Braun's team of engineers were engaged in the development of large rockets. Eventually NASA created other Centers and today it has ten located around the country.

NASA began to conduct space missions within months of its creation, and during its first twenty years NASA conducted several major programs:

- Human space flight initiatives—Mercury's single astronaut program (flights during 1961–1963) to ascertain if a human could survive in space; Project Gemini (flights during 1965–1966) with two astronauts to practice space operations, especially rendezvous and docking of spacecraft and extravehicular activity (EVA); and Project Apollo (flights during 1968–1972) to explore the Moon.
- Robotic missions to the Moon (Ranger, Surveyor, and Lunar Orbiter), Venus (*Pioneer Venus*), Mars (*Mariner 4* and *Viking 1* and *2*), and the outer planets (*Pioneer 10* and *11* and *Voyager 1* and *2*).
- Aeronautics research to enhance air transport safety, reliability, efficiency, and speed (X-15 hypersonic flight, lifting body flight research, avionics and electronics studies, propulsion technologies, structures research, aerodynamics investigations).
- Remote-sensing Earth satellites for information gathering (Landsat satellites for environmental monitoring).
- Applications satellites for communications (*Echo 1*, *TIROS*, and *Telstar*) and weather monitoring.
- An orbital workshop for astronauts, *Skylab*.
- A reusable spacecraft for traveling to and from Earth orbit, the Space Shuttle.

Early Spaceflights: Mercury and Gemini

NASA's first high-profile program involving human spaceflight was Project Mercury, an effort to learn if humans could survive the rigors of spaceflight. On May 5, 1961, Alan B. Shepard, Jr. became the first American to fly into space, when he rode his Mercury capsule on a 15-minute suborbital mission. John H. Glenn Jr. became the first U.S. astronaut to orbit the Earth on February 20, 1962. With six flights, Project Mercury achieved its goal of putting piloted spacecraft into Earth orbit and retrieving the astronauts safely.

Project Gemini built on Mercury's achievements and extended NASA's human space flight program to spacecraft built for two astronauts. Gemini's ten flights also provided NASA scientists and engineers with more data on weightlessness, perfected re-entry and splashdown procedures, and demonstrated rendezvous and docking in space. One of the highlights of the program occurred during *Gemini 4*, on June 3, 1965, when Edward H. White, Jr., became the first U.S. astronaut to conduct a space walk.

Going to the Moon—Project Apollo

The singular achievement of NASA during its early years involved the human exploration of the Moon, Project Apollo. Apollo became a NASA priority on May 25, 1961, when President John F. Kennedy announced "I believe that this nation should commit itself to achieving the goal, before this decade is out, of landing a man on the Moon and returning him safely to Earth." A direct response to Soviet successes in space, Kennedy used Apollo as a high-profile effort for the United States to demonstrate to the world its scientific and technological superiority over its cold war adversary.

In response to the Kennedy decision, NASA was consumed with carrying out Project Apollo and spent the next 11 years doing so. This effort required significant expenditures— $25.4 billion over the life of the program. Only the building of the Panama Canal rivaled the size of the Apollo program as the largest nonmilitary technological endeavor ever undertaken by the United States (the Manhattan Project was comparable in a wartime setting). Although there were major challenges and some failures—notably a January 27, 1967, fire in an Apollo capsule on the ground that took the lives of astronauts Roger B. Chaffee, Virgil "Gus" Grissom, and Edward H. White, Jr.—the program moved forward inexorably.

Less than two years later, in October 1968, NASA bounced back with the successful *Apollo 7* mission, which orbited the Earth and tested the redesigned Apollo command module. The *Apollo 8* mission, which orbited the Moon on December 24–25, 1968 while its crew read from the book of Genesis, was another crucial accomplishment on the way to the Moon.

"That's one small step for [a] man, one giant leap for mankind." Neil A. Armstrong uttered these famous words on July 20, 1969, when the *Apollo*

11 mission fulfilled Kennedy's challenge by successfully landing Armstrong and Edwin E. "Buzz" Aldrin, Jr. on the Moon. Armstrong dramatically piloted the lunar module to the lunar surface with less than 30 seconds worth of fuel remaining. After taking soil samples, photographs, and performing other tasks on the Moon, Armstrong and Aldrin rendezvoused with their colleague Michael Collins in lunar orbit for a safe voyage back to Earth.

Five more successful lunar landing missions followed. The *Apollo 13* mission of April 1970 attracted the public's attention when astronauts and ground crews had to improvise to end the mission safely after an oxygen tank burst midway through the journey to the Moon. Although this mission never landed on the Moon, it reinforced the notion that NASA had a remarkable ability to adapt to the unforeseen technical difficulties inherent in human space flight.

With the *Apollo 17* mission of December 1972, NASA completed a successful engineering and scientific program. Fittingly, Harrison H. "Jack" Schmitt, a geologist who participated on this mission, was the first scientist to be selected as an astronaut. NASA learned a good deal about the origins of the Moon, as well as how to support humans in outer space. In total, tweleve astronauts walked on the Moon during six Apollo lunar landing missions.

In 1975, NASA cooperated with the Soviet Union to achieve the first international human space flight, the Apollo-Soyuz Test Project (ASTP). This project successfully tested joint rendezvous and docking procedures for spacecraft from the U.S. and the U.S.S.R. After being launched separately from their respective countries, the Apollo and Soyuz crews met in space and conducted various experiments for two days.

Space Shuttle

After a gap of six years, NASA returned to human space flight in 1981, with the advent of the Space Shuttle. The Shuttle's first mission, *STS-1*, took off on April 12, 1981, demonstrating that it could take off vertically and glide to an unpowered airplane-like landing. On *STS-6*, during April 4–9, 1983, F. Story Musgrave and Donald H. Peterson conducted the first Shuttle EVA, to test new spacesuits and work in the Shuttle's cargo bay. Sally K. Ride became the first American woman to fly in space when *STS-7* lifted off on June 18, 1983, another early milestone of the Shuttle program.

On January 28, 1986, a leak in the joints of one of two Solid Rocket Boosters attached to the *Challenger* orbiter caused the main liquid fuel tank to explode 73 seconds after launch, killing all seven crewmembers. The Shuttle program was grounded for over two years, while NASA and its contractors worked to redesign the solid rocket boosters and implement management reforms to increase safety.

Tragedy struck again on February 1, 2003, when the *Columbia* orbiter broke up during reentry just minutes away from landing. The investigation into the cause of the crash was expected to be completed in the summer of 2003. The leading theory was that a chunk of insulating foam broke off during take off and created a gash in the leading edge of the shuttle wing leaving a portion of the shuttle exposed to the extreme temperatures experienced during reentry.

On June 6, the Columbia Accident Investigation board tried to recreate the damage by firing a block of insulating foam at a left wing panel. The foam created a three-inch crack and investigators also found a crack in the seal between panels of the wing. The experiment bolstered the theory that the damage caused by the foam triggered the accident.

In June, a NASA "return-to-flight" team announced several changes it would make to launch protocol and the remaining Shuttle orbiters—*Atlantis*, *Discovery*, and *Endeavour*—to improve safety. The team hoped to resume Shuttle launches as early as December 2003.

Toward a Permanent Human Presence in Space

The core mission of any future space exploration will be humanity's departure from Earth orbit and journeying to the Moon or Mars, this time for extended and perhaps permanent stays. A dream for centuries, active efforts to develop both the technology and the scientific knowledge necessary to carry this off are now well underway. The next generation of launch vehicles taking us from the Earth into orbit is being developed right now. The X-33, X-34, and other hypersonic research projects presently underway will help to realize routine, affordable access to space in the first decades of the twenty-first century.

An initial effort in this area was NASA's Skylab program in 1973. After Apollo, NASA used its huge Saturn rockets to launch a relatively small orbital space workshop. There were three human Skylab missions, with the crews staying aboard the orbital workshop for 28, 59, and then 84 days. The first crew manually fixed a broken meteoroid shield, demonstrating that humans could successfully work in space. The Skylab program also served as a successful experiment in long-duration human space flight.

In 1984, Congress authorized NASA to build a major new space station as a base for further exploration of space. By 1986, the design depicted a complex, large, and multipurpose facility. In 1991, after much debate over the station's purpose and budget, NASA released plans for a restructured facility called Space Station Freedom. Another redesign took place after the Clinton administration took office in 1993 and the facility became known as Space Station Alpha.

Then Russia, which had many years of experience in long-duration human space flight, such as with its *Salyut* and *Mir* space stations, joined with the U.S. and other international partners in 1993 to build a joint facility that became known formally as the International Space Station (ISS). To prepare for building the ISS starting in late 1998, NASA participated in a series of Shuttle missions to *Mir* and seven American astronauts lived aboard *Mir* for extended stays.

The Science of Space

In addition to major human space flight programs, there have been significant scientific probes that have explored the Moon, the planets, and other areas of our solar system. In particular, the 1970s heralded the advent of a new generation of scientific spacecraft. Two similar spacecraft, *Pioneer 10* and *Pioneer 11*, launched on March 2, 1972, and April 5, 1973, respectively, traveled to Jupiter and Saturn to study the composition of interplanetary space. *Voyagers 1* and *2*, launched on September 5, 1977, and August 20, 1977, respectively, conducted a "Grand Tour" of our solar system.

In 1990, the Hubble Space Telescope (HST) was launched into orbit around the Earth. Unfortunately, NASA scientists soon discovered that a microscopic spherical aberration in the polishing of the Hubble's mirror significantly limited the instrument's observing power. During a previously scheduled servicing mission in December 1993, a team of astronauts performed a dramatic series of space walks to install a corrective optics package and other hardware. The hardware functioned like a contact lens and the elegant solution worked perfectly to restore Hubble's capabilities. The servicing mission again demonstrated the unique ability of humans to work in space, enabled Hubble to make a number of important astronomical discoveries, and greatly restored public confidence in NASA.

Several months before this first HST servicing mission, however, NASA suffered another major disappointment when the *Mars Observer* spacecraft disappeared on August 21, 1993, just three days before it was to go into orbit around the red planet. In response, NASA began developing a series of better, faster, cheaper spacecraft to go to Mars.

Mars Global Surveyor was the first of these spacecraft; it was launched on November 7, 1996, and has been in a Martian orbit mapping Mars since 1998. Using some innovative technologies, the *Mars Pathfinder* spacecraft landed on Mars on July 4, 1997, and explored the surface of the planet with its miniature rover, Sojourner. The *Mars Pathfinder* mission was a scientific and popular success, with the world following along via the Internet.

Over the years, NASA has continued to look for life beyond our planet. In 1975, NASA launched the two Viking spacecraft to look for basic signs of life on Mars; the spacecraft arrived on Mars in 1976 but did not find any indications of past or present biological activity there. In 1996 a probe from the *Galileo* spacecraft that was examining Jupiter and its moon, Europa, revealed that Europa may contain ice or even liquid water, thought to be a key component in any life-sustaining environment. NASA also has used radio astronomy to scan the heavens for potential signals from extraterrestrial intelligent life. It continues to investigate whether any Martian meteorites contain microbiological organisms and in the late 1990s, organized an Origins program to search for life using powerful new telescopes and biological techniques.

The "First A in NASA": Aeronautics Research

Building on its roots in the National Advisory Committee for Aeronautics, NASA has continued to conduct many types of cutting-edge aeronautics research on aerodynamics, wind shear, and other important topics using wind tunnels, flight-testing, and computer simulations. In the 1960s, NASA's highly successful X-15 program involved a rocket-powered airplane that flew above the atmosphere and then glided back to Earth unpowered. The X-15 pilots helped researchers gain much useful information about supersonic aeronautics and the program also provided data for development of the Space Shuttle. NASA also cooperated with the air force in the 1960s on the X-20 Dyna-Soar program, which was designed to fly into orbit. The Dyna-Soar was a precursor to later, similar efforts such as the National Aerospace Plane, on which NASA and other government agencies and private companies did advanced hypersonics research in such areas as structures, materials, propulsion, and aerodynamics.

NASA has also done significant research on flight maneuverability on high-speed aircraft that is often applicable to lower-speed airplanes. NASA scientist Richard Whitcomb invented the supercritical wing that was specially shaped to delay and lessen the impact of shock waves on transonic military aircraft; it had a significant impact on civil aircraft design. Beginning in 1972, the watershed F-8 digital-fly-by-wire (DFBW) program laid the groundwork for electronic DFBW flight in various later aircraft such as the F/A-18, the Boeing 777, and the Space Shuttle. More sophisticated DFBW systems were used on the X-29 and X-31 aircraft, which would have been uncontrollable otherwise.

From 1963 to 1975, NASA conducted a research program on lifting bodies, aircraft without wings. This valuable research paved the way for the Shuttle as well as for the later X-33 project to glide to a safe unpowered landing and for a prototype for a future crew return vehicle from the International Space Station.

Applications Satellites

NASA did pioneering work in space applications such as communications satellites in the 1960s. The Echo, Telstar, Relay, and Syncom satellites were built by NASA or by the private sector based on significant NASA advances.

In the 1970s, NASA's Landsat program literally changed the way we look at our planet Earth. The first three Landsat satellites, launched in 1972, 1975, and 1978, transmitted back to Earth complex data streams that could be converted into colored pictures. Landsat data has been used in a variety of practical commercial applications such as crop management and fault line detection, and to track many kinds of weather including droughts, forest fires, and ice floes. NASA has been involved in a variety of other Earth science efforts such as the Earth Observation System of spacecraft and data processing that have yielded important scientific results in such areas as tropical deforestation, global warming, and climate change.

Conclusion

Since its inception in 1958, NASA has accomplished many great scientific and technological feats. NASA technology has been adapted for many nonaerospace uses by the private sector. Today, NASA remains a leading force in scientific research and in stimulating public interest in aerospace exploration, as well as in science and technology in general. Perhaps more importantly, our exploration of space has taught us to view the Earth, ourselves, and the universe in a new way. While the tremendous technical and scientific accomplishments of NASA demonstrate vividly that humans can achieve previously inconceivable feats, we also are humbled by the realization that Earth is just a tiny blue marble in the cosmos.

Bibliography

Bilstein, Roger E., *Orders of Magnitude: A History of the NACA and NASA, 1915–1990.* NASA SP-4406 (Washington, D.C.: Government Printing Office, 1989).

Columbia Accident Investigation Board. http://www.caib.us. Accessed 11 June 2003.

Logsdon, John M., Linda J. Lear, Jannelle Warren Findley, Ray A. Williamson, and Dwayne A. Day. *Organizing for Exploration*, vol 1 of *Exploring the Unknown: Selected Documents in the History of the U.S. Civil Space Program*. NASA SP-4407. (Washington, D.C.: Government Printing Office, 1995).

Logsdon, John M., Dwayne A. Day, and Roger D. Launius. *External Relationships*, vol. 2 of *Exploring the Unknown: Selected Documents in the History of the U.S. Civil Space Program*. NASA SP-4407. (Washington, D.C.: Government Printing Office, 1996).

Logsdon, John M., Roger D. Launius, David H. Onkst, and Stephen J. Garber. *Using Space*, vol. 3 of *Exploring the Unknown: Selected Documents in the History of the U.S. Civil Space Program*. NASA SP-4407. (Washington, D.C.: Government Printing Office, 1998).

Reinert, Patty, "NASA Eyes December for Next Shuttle Flight." *Houston Chronicle,* 11 June 2003, A1.

Appendix B

Regional Technology Transfer Centers

The Robert C. Byrd National Technology Transfer Center

Wheeling Jesuit University
316 Washington Ave.
Wheeling, WV 26003
(800) 678-6882

Since its establishment in 1990, the National Technology Transfer Center (NTTC) has acted on more than 10,000 technical requests and recorded more than 100,000 log-ons to its electronic information system.

Funded by NASA, and with additional support from the Ballistic Missile Defense Organization, the U.S. Navy, and other federal agencies, the NTTC has its headquarters on the campus of Wheeling Jesuit College in Wheeling, West Virginia. From there it extends its reach throughout the country, drawing on the expertise of more than 700 laboratories and thousands of scientists, engineers, and managers who make up the Federal Laboratory Consortium. NTTC also teams with NASA's Regional Technology Transfer Centers (RTTCs)—six offices nationwide that assist local industry in technology evaluation and commercialization, as well as market assessment and other areas of business expertise.

NTTC's programs run the gamut of technology transfer initiatives. A key service is called the National Gateway, providing callers from business and industry free access to technology and expertise from the NASA labs and other federally funded research facilities. This is done through NTTC technology access agents, who, once apprised of a technical challenge or area of interest, search the proper databases, contact the labs, and often put the inquirer in touch with the individual experts who are the best resources.

But if the inquirer simply wants to browse through the ideas and technical information available from federal sources, there is Business Gold, a free service available via the Internet or direct modem contact. Operating 24 hours a day, NTTC's electronic bulletin board provides information that can be downloaded onto the user's computer free of all charges except for the phone lines.

Included are descriptions of new technologies available for licensing and commercialization; NASA RTTC contacts; current SBIR, Technology Reinvestment Program, and similar solicitations; technology transfer conference calendars; information from government software centers; and more.

Education and training are significant aspects of NTTC's activities. Idea Contact, a new program directed at federal lab and agency personnel, teaches scientists, engineers, and managers how to evaluate a technical innovation for its manufacturability, marketability, and usefulness to the commercial world. For small- and medium-sized manufacturers, NTTC provides technology-driven industrial extension programs, concentrating on the use of federal technologies as a catalyst for economic revitalization. As part of the Technology Reinvestment Program, NTTC received a grant to work with the Pennsylvania Technical Assistance Program, Rensselaer Polytechnic Institute, and other manufacturing assistance groups to create a national curriculum for training industrial extension agents. Finally, in cooperation with Wheeling Jesuit College, NTTC has developed courses of study for bachelor and master of science degrees in technology management, innovation, commercialization, and technology transfer.

NASA's Regional Technology Transfer Centers

"A network within a network." That is how the staff at the Center for Technology Commercialization aptly describes its role as one of the six NASA Regional Technology Transfer Centers. Each of them reaches out to its national brethren, as well as to the hub, the National Technology Transfer Center at Wheeling Jesuit College, Wheeling, West Virginia. But at the same time, each, in ways that are similar but uniquely suited to its locale, stands at the center of a regional network linking federal laboratories with state and local agencies, universities, and businesses.

The RTTCs, funded by NASA and aligned with the six Federal Laboratory Consortium regions, opened their doors in January 1992. Since then they have established regional ties to more than 70 state and local organizations, creating a national web to enable U.S. companies to learn of, evaluate, and acquire NASA and other federally funded technologies for commercial exploitation.

Center for Technology Commercialization (CTC)

1400 Computer Drive
Westborough, MA 01881-5043
(508) 870-0042
Web site: www.ctc.org

CTC, serving the Northeast from Westborough, Massachusetts, defines its mission as making American companies more competitive worldwide through market-driven product and process definition, technology acquisition, and product commercialization. The corporation's "inner" network comprises eight Satellite Technology Transfer Centers located in the six New England states, New York, and New Jersey. These are responsible for knowing industrial market needs and capabilities, opportunities, and programs throughout their local areas.

CTC's NASA Business Outreach Program, established in 1993, acts as advocate for Northeast businesses in seeking contracting opportunities with NASA and its prime contractors. The program focuses on small, minority-owned, and woman-owned firms to help them realize the full potential of NASA opportunities for their products, services, and technology areas.

CTC also participates in the Advanced Research Projects Agency's TAP-IN, a program that helps small defense-related companies develop commercial products, enter new markets, and find commercial applications for defense technology. Other resources include CTC's Technical Information Center, which facilitates technology, marketing, and patent research, and document and patent delivery. (Source: *NASA Spinoff 2001*)

The Technology Commercialization Center (TeCC)

12050 Jefferson Avenue, Suite 340
Newport News, VA 23606
(757) 269-0025
Web site: www.teccenter.org

TeCC operates the NASA Regional Technology Commercialization Center for the Mid-Atlantic Region, which comprises Delaware, Maryland, Pennsylvania, Virginia, and West Virginia as part of the NASA Commercial Technology Network.

Our goal is to encourage industries in the Mid-Atlantic Region, as well as nationwide, to commercialize NASA-developed technologies. Our range of technology transfer and commercialization services includes technology evaluation, technology marketing, technology needs assessments, and information services.

We analyze and package NASA technologies, market the packages through state field agents, match industry with technologies, and assist industry with steps to bring technology to the market.

Our Web site (www.teccenter.org) describes our marketing effort, provides information on available technologies, explains the commercialization process, and invites companies to register for additional information. This Web site is integrated into the NASA technology commercialization network.

TeCC was incorporated in Virginia as a non-profit corporation in 1999 and our staff is organized around the NASA technologies sectors that offer the highest potential for commercial opportunities: advanced materials;

sensors and instrumentation; mechanical devices; transportation applications; medical applications. (Source: TeCC Web site)

Southeast Regional Technology Transfer Center (SERTIC)

216 O'Keefe Building
Atlanta, GA 30332-0640
(800) 472-6785
Web site: www.edi.gatech.edu/nasa/

Global communication has flourished in the past decade, uniting the world's markets and driving increased competition. To remain competitive, especially in the high technology arena, companies must create alliances to limit technology costs, reduce time-to-market, and increase product variability. To help your firm overcome these competitive hurdles, NASA has established RRTTCs to facilitate partnerships that can connect industry's technology needs with NASA's technology solutions.

In the Southeast, the RTTC is part of the Georgia Institute of Technology (Georgia Tech). The SERTTC is funded by NASA to:

- Spread the word to industry that NASA has a wealth of technology transfer opportunities, including licensing and partnering activities.
- Assist companies in finding the appropriate NASA technology opportunities that best suit their product development needs.
- Guide companies through the NASA technology transfer process.

The SERTTCs supports three NASA Field Centers: the Kennedy Space Center in Florida, Marshall Space Flight Center in Alabama, and John C. Stennis Space Center in Mississippi. This is done through the combined efforts of a nine state affiliate network. Administratively, the SERTTC is part of the Georgia Institute of Technology's VentureLab. VentureLab is part of the Economic Development and Technology Ventures group, which houses the Advanced Technology Development Center (ATDC) and the Economic Development Institute (EDI). (Source: SERTTC Web site)

Great Lakes Industrial Technology Center (GLITeC)

20445 Emerald Parkway Drive, S.W.
Suite 200
Cleveland, OH 44135
(216) 898-6400
Web site: www.glitec.org

What can space-related research do for your Earth-bound enterprise? Everything.

NASA success has been as far ranging and creative as you can imagine. Work aimed at landing rovers on Mars has been used in safeguarding the highways here at home. Advanced antenna research, performed to improve space stations and spacecraft, has bolstered the cellular and satellite communications market. NASA researchers have even worked with companies who may use Glenn innovations to design new children's toys. Now, your company could be NASA's next big success story, thanks to GLITeC.

GLITeC and its affiliates work with over five hundred companies a year, with needs ranging from locating technical experts and information to identifying new market and product opportunities. GLITeC's services are designed to help companies identify, acquire, adapt, and use federal technology and capabilities. We serve a variety of companies, of all sizes and in all areas of technology.

GLITeC draws on NASA and federal laboratories, state and federal technology application centers, and university centers for technical expertise in order to offer information services, technology needs assessment, and commercialization services.

Information Services

GLITeC's information services start with computerized data searches for useful technology and funding possibilities. Also included are NASA patent searches and NASA technical reports.

Technology Needs Assessment

GLITeC can also assess your company's technology needs. GLITeC staff or NASA scientists will analyze your technology needs and provide technical consulting, brainstorming, or referral services.

Commercialization Services

GLITeC provides a number of commercialization services. These services range from reviewing business plans to introducing our clients to venture capitalists. GLITeC can also identify and facilitate access to NASA researchers or to NASA consulting. GLITeC staff helps companies prepare to license NASA patents and develops cooperative joint development projects to support new product introductions or process improvements.

GLITeC is managed by the Battelle Memorial Institute, the world's largest not-for-profit industrial R&D organization with:

- Nearly $1 billion in annual business volume

- 7,500 staff dedicated to technology development, commercialization, and management
- 70 years experience servicing industry
- 5,000 current projects for 1,600 government and industry clients in 50 countries. (Source: GLiTeC Web site)

Mid-Continent Technology Transfer Center (MCTTC)

301 Tarrow
College Station, TX 77843-8000
(409) 845-8762
Web site: www.mcttc.com

As one of six NASA-funded RTTCs, the MCTTC offers a variety of technology transfer and commercialization services. It serves private companies and federal laboratories by forging a value-added link between technology sources and recipients.

The MCTTC provides this assistance to a 14-state region from its headquarters in College Station, Texas, and through a team of subcontractors and affiliates. The team, composed of universities, private industry, and federal and state agencies, caters primarily to companies in the Continent region, but it pulls essential technologies from more than 600 laboratories nationwide. The MCTTC reflects NASA's initiative upgrade and restructures its technology transfer program to better serve U.S. business and industry in the 21st century and beyond.

The MCTTC is under the direction of the Technology and Economic Development Division of the Texas Engineering Extension Service (TEEX), a member of the Texas A&M University System. The MCTTC reports directly to NASA Johnson Space Center, and it is a participating member of the NASA Commerical Technology Network. (Source: MCTTC Web site)

Far West Regional Technology Transfer Center

University of Southern California
3716 S. Hope St., Suite 200
Los Angeles, CA 90007-4344
(213) 743-2353
Web site: www.usc.edu/dept/engineering/TTC/NASA/index.html

Headquartered at the University of Southern California in Los Angeles, the Far West Regional Technology Transfer Center (Far West RTTC) serves an eight-state area: Arizona, California, Nevada, Washington, Oregon, Idaho, Alaska, and Hawaii. One of six NASA-sponsored technology transfer centers across the country, the Far West RTTC is an Engineering Research Center within the School of Engineering at the University of Southern California.

Appendix C

NASA Field Centers and Facilities

NASA Headquarters

300 E Street SW
Washington, DC 20024-3210
(202) 358-0000
Web site: www.nasa.gov

NASA Headquarters, located in Washington, D.C., exercises management over the space flight centers, research centers, and other installations that constitute NASA. Responsibilities of Headquarters cover the determination of programs and projects; establishment of management policies, procedures, and performance criteria; evaluation of progress; and the review and analysis of all phases of the aerospace program.

Ames Research Center (ARC)

Moffett Field, CA 94035
(650) 604-5000
Web site: www.arc.nasa.gov
Technology transfer Web site: http://ctoserver.arc.nasa.gov/

NASA ARC is located at Moffett Field, California. Ames is a principal center for computational fluid dynamics, rotorcraft and powered-lift technology, artificial intelligence, and airborne sciences. Other specialties include flight simulation, robotics, human factors research, advanced life support research, pathfinding research in the fundamental biological sciences, origin of life and exobiology research, and wind tunnel design, development, and operation.

Selected technological strengths: Fluid Dynamics; Life Sciences; Earth and Atmospheric Sciences; Information, Communications, and Intelligent Systems; Human Factors.

Dryden Flight Research Center (DFRC)

P.O. Box 273
Edwards, CA 93523-0273
(661) 276-3311
Web site: www.dfrc.nasa.gov
Technology transfer Web site: www.dfrc.nasa.gov/TechTransfer
/TechTransfer.html

DFRC is NASA's premier installation for aeronautical flight research. It is located at Edwards Air Force Base (AFB), California. Dryden's work spans the early jet and space age, from breaking the sound barrier to flying the Space Shuttle.

Selected technological strengths: Aerodynamics; Aeronautics Flight Testing; Aeropropulsion; Flight Systems; Thermal Testing; Integrated Systems Test and Validation.

Glenn Research Center (GRC)

Lewis Field
21000 Brookpark Road
Cleveland, OH 44135
(216) 433-4000
Web site: www.grc.nasa.gov
Technology transfer Web site: http://cto.grc.nasa.gov/

NASA has designated GRC as its Lead Center for Aeropropulsion, which is located in Cleveland, Ohio. Their role is to develop, verify, and transfer aeropropulsion technologies to U.S. industry. As NASA's designated Center of Excellence in Turbomachinery, their role is to develop new and innovative turbomachinery technology to improve the reliability and performance, efficiency and affordability, capacity and environmental compatibility of future aerospace vehicles.

Selected technological strengths: Aeropropulsion; Communications; Energy Technology; High Temperature Materials Research.

Goddard Institute for Space Studies (GISS)

2880 Broadway
New York, NY 10025
(212) 678-5500
Web site: www.giss.nasa.gov

GISS is a NASA research institute located near Columbia University in New York City. GISS is a division of Goddard Space Flight Center's Earth Sciences Directorate. GISS is primarily engaged in studies of global climate change.

Goddard Space Flight Center (GSFC)

Greenbelt, MD 20771
(301) 286-8955
Web site: www.gsfc.nasa.gov
Technology transfer Web site: http://techtransfer.gsfc.nasa.gov/

NASA's Center of Excellence for Scientific Research. GSFC is located in suburban Maryland, northeast of Washington, DC. This NASA field center is a major U.S. laboratory for developing and operating unmanned scientific spacecraft. GSFC manages many of NASA's Earth Observation, Astronomy, and Space Physics missions.

Selected technological strengths: Earth and Planetary Science Missions; LIDAR; Cryogenic Systems; Tracking; Telemetry; Command.

Independent Verification & Validation Facility

100 University Drive
Fairmont, WV 26554
(304) 367-8200
Web site: www.ivv.nasa.gov/

The NASA Software Independent Verification & Validation Facility was established in 1993 as part of an agency-wide strategy to provide the highest achievable levels of safety and cost-effectiveness for mission critical software. The IV&V Facility was founded under the NASA Office of Safety and Mission Assurance as a direct result of recommendations made by the National Research Council (NRC) and the Report of the Presidential Commission on the Space Shuttle Challenger Accident.

Jet Propulsion Laboratory (JPL)

4800 Oak Grove Drive
Pasadena, CA 91109
(818) 354-4321
Web site: www.jpl.nasa.gov
Technology transfer Web site: http://techtransfer.jpl.nasa.gov/

NASA's Center of Excellence for Deep Space Systems. Managed for NASA by the California Institute of Technology, JPL is the lead U.S. Center for robotic exploration of the solar system. Located in Pasadena, California, JPL focuses efforts on planetary exploration (including Galileo, Voyager, Magellan and missions to Mars), and environmental research (including the Shuttle Imaging Radar and TOPEX/POSEIDON).

Selected technological strengths: Near/Deep-Space Mission Engineering; Microspacecraft; Space Communications; Information Systems; Remote Sensing; Robotics.

Johnson Space Center (JSC)

2101 NASA Road 1
Houston, TX 77058-3696
(281) 483-0123
Web site: www.jsc.nasa.gov
Technology transfer Web site: http://technology.jsc.nasa.gov/

NASA's Center for Excellence for Human Operations in Space. Lyndon B. Johnson Space Center is located in Houston, Texas. JSC was established in September 1961 as NASA's primary center for design, development and testing of spacecraft and associated systems for Human Flight; Selection and Training of Astronauts; and Planning and Conducting Human Space Flight Missions.

Selected technological strengths: Artificial Intelligence and Human Computer Interface; Life Sciences; Human Space Flight Operations; Avionics; Sensors; Communications.

Kennedy Space Center (KSC)

Kennedy Space Center, FL 32899
(321) 867-5000
Web site: www.ksc.nasa.gov
Technology transfer Web site: http://technology.ksc.nasa.gov/

NASA's Center of Excellence for Launch and Cargo Processing Systems. The John F. Kennedy Space Center (KSC) has primary responsibility for ground turnaround and support operations, prelaunch checkout and launch of the Space Shuttle and its payloads, including NASA's International Space Station. KSC is located in Florida, and is the nation's spaceport, the liftoff site for all manned missions into space.

Selected technological strengths: Environmental Monitoring; Sensors; Corrosion Protection; Bio-Sciences; Process Modeling; Work Planning/Control; Meteorology.

Langley Research Center (LaRC)

100 NASA Road
Hampton, VA 23681-2199
(757) 864-1000
Web site: www.larc.nasa.gov
Technology transfer Web site: http://tech-transfer.larc.nasa.gov/

NASA's Center of Excellence for Structure and Materials. LaRC is located in Hampton, Virginia and its primary mission is basic research in aeronautics and space technology. LaRC is lead center for management of the Agency's technology development programs for future high-speed civil transport, for hypersonic vehicle concepts, and for general aviation.

Selected technological strengths: Aerodynamics; Flight Systems; Materials; Structures; Sensors; Measurements; Information Sciences.

Marshall Space Flight Center (MSFC)

4200 Rideout Road Bldg 4200
Huntsville, AL 35806
(256) 544-4524
Web site: www1.msfc.nasa.gov
Technology transfer Web site: http://techtran.msfc.nasa.gov/

The George C. Marshall Space Flight Center in Huntsville, Alabama, is NASA's premier organization for developing space transportation and propulsion systems and for conducting microgravity research. MSFC is also NASA's Center of Excellence in Space Propulsion and Space Optical Systems.

Selected technological strengths: Materials; Manufacturing; Nondestructive Evaluation; Biotechnology; Space Propulsion; Controls and Dynamics; Structures; Microgravity Processing.

Stennis Space Center (SSC)

Bay St. Louis, MS 39529
(228) 688-3390
Web site: www.ssc.nasa.gov
Technology transfer Web site: http://technology.ssc.nasa.gov/

The John C. Stennis Space Center is NASA's primary center for testing large rocket propulsion systems for the Space Shuttle and future generation space vehicles. Because of its important role in engine and vehicle testing over the past three decades, SSC has been designated NASA's Center of Excellence for Large Propulsion Systems Testing. SSC is located in Hancock County, Mississippi.

Selected technological strengths: Propulsion Systems; Test/Monitoring; Remote Sensing; Nonintrusive Instrumentation.

Wallops Flight Facility (WFF)

Wallops Island, VA 23337
(757) 824-1579
Web site: wff.nasa.gov

WFF is located at Wallops Island, Virginia, and is managed by the Goddard Space Flight Center, in Greenbelt, Maryland. Wallops, one of the oldest launch sites in the world, has over the years been meeting the needs of the United States aerospace program. A sizable portion of Wallops' effort is devoted to aeronautical research and development and in reporting the nation's space technology to the international community.

(Reprinted courtesy NASA centers, Spacelink and NASA Tech)

Appendix D

NASA's Space Product Development Program

The commercial development of the space frontier is one of the greatest opportunities facing America. It is the growth of business into space that will bring the benefits of space down to Earth and enrich the everyday lives of all Americans. NASA is encouraging businesses to seize this opportunity through its Space Product Development Office, to ensure the continued economic growth of the United States and to bring the opportunities for new advances, technological understanding, products, and jobs to the public.

This effort is one of NASA's major goals, and the goal of the Space Product Development Office is to help American business explore the potential—and reap the rewards—of doing business in space. This not only ensures improvements to our everyday lives, but also establishes a commercial demand for space. Doing this, however, requires that seeds be planted with American businesses.

These seeds are in the form of alliances with industry and academia through Commercial Space Centers that support the full spectrum of commercial research. These Centers, located at academic institutions such as universities, are currently funded by NASA and charged with developing industry partners to pursue specific areas of commercial research. These partners have to help pay an increasing portion of the funding for operations and research, since the ultimate goal is to generate a demand for doing business in space. (Source: Space Product Development Web site. http://spd.nasa.gov/)

Commercial Space Centers

Commercial Space Centers, located at academic institutions and funded by NASA, develop industry partners to pursue specific areas of commercial research. The industry partners generate a demand for doing business in space and help fund a portion of the operations and research.

The Centers are subject to several types of review and appraisal. One measure used is the interest of industry, which is evaluated by the number of industry partners and funding. Another measure checks progress against previously established milestones. Periodic checks are made by an outside source, such as the National Academy of Public Administration, to provide an independent assessment of the Centers and their performance.

Currently, NASA sponsors seventeen Commercial Space Centers, with fifteen of them operating through the Space Product Development Program. As new areas of commercial interest are identified, new Centers will be established, while other Centers may be disbanded if industry involvement wanes. The current Commercial Space Centers are:

BioServe Space Technologies

University of Colorado—Boulder
Aerospace Engineering Sciences
429 UCB
Boulder, CO 80309-0429
(303) 492-4010
Web sites: http://www.colorado.edu/engineering/BioServe
http://www.ksu.edu/bioserve/
Focus: Provides research support and specialized hardware for space flight in the following areas: pharmaceuticals, biotechnology, biomedical, agriculture.

Center for Advanced Microgravity Materials Processing (CAMMP)

Department of Chemical Engineering
342 Snell Engineering
Boston, MA 02115
(617) 373-7910
Web sites: http://www.dac.neu.edu/cammp/
Focus: Improves the production of advanced catalytic materials for industry to include the following: zeolite crystals, ferroelectrics.

Center for Biophysical Science and Engineering (CBSE)

Center for Biophysical Science and Engineering
University of Alabama at Birmingham
MCLM 262
1530 3rd Avenue South
Birmingham, AL 35294-0005
(205) 934-5329
Web site: http://www.cbse.uab.edu/
Focus: Determines the structures of proteins for pharmaceutical research.

Center for Commercial Applications of Combustion in Space (CCACS)

CCACS
Colorado School of Mines
1500 Illinois St.
Golden, CO 80401-1887
(303) 384-2091
Web site: http://www.mines.edu/research/ccacs/
Focus: Helps industry improve production and safety by understanding combustion for the following processes: chemical, manufacturing.

Center for Satellite and Hybrid Communication Networks Institute for Systems Research

University of Maryland
A.V. Williams Building 115
College Park, MD 20742
(301) 405-7900
Web site: http://www.isr.umd.edu/CSHCN/
Focus: Develops hybrid networks that link satellite and wireless systems with the following networks: cellular, cable, Internet, telephone.

Center for Space Power

Texas A&M University
Wisenbaker Building, Room 223
College Station, TX, 77843-3118
(979) 845-8768
Web site: http://engineer.tamu.edu/tees/csp/
Focus: Develops a variety of space power-related technologies to include the following: loop heat pipes, lithium batteries, ilmenite semiconductor materials, microwave electrothermal thruster, digital communications algorithms, magnetic bearing control algorithms, high-efficiency power conditions.

Center for Space Power and Advanced Electronics

Space Research Institute
231 Leach Center
Auburn University, AL 36849
(334) 844-5894
Web site: http://spi.auburn.edu/

Focus: Advances technology associated with the following: high-temperature SiC devices, high-temperature electronics and packaging.

Commercial Space Center for Engineering (CSCE)

223 Weisenbaker Engineering Research Center
MS3118
Texas A&M University
College Station, TX 77843-3118
(979) 845-8768
http://engineer.tamu.edu/tees/csce/
Focus: Provides a testbed on the International Space Station (ISS) for developing advanced spacecraft technology in the following areas: solar arrays, antennas, sensors, other satellite components.

Consortium for Materials Development in Space (CMDS)

Consortium for Materials Development in Space
University of Alabama in Huntsville
Research Institute Building
4701 University Drive
Huntsville, AL 35899
(256) 824-6620
http://www.uah.edu/research/cmds/index.html
Focus: Develops new and improved materials in space and on the ground to include the following: organic and inorganic matter, space structures, advancements in medicines.

Environmental Systems Commercial Space Technology Center

P.O. Box 116450
University of Florida
Gainesville, FL 32611-6450
(352) 392-7814
Web site: http://www.ees.ufl.edu/escstc/
Focus: Develops technologies to be used in long-duration human space flight, while at the same time applying those technologies in the commercial sector.

Food Technology Commercial Space Center

Iowa State University
2901 South Loop Drive
Suite 3700
Ames, IA 50010-8632
(515) 296-5383
Focus: Develops food technology so astronauts will have a food supply from crops grown and harvested in space and on earth.

Medical Informatics and Technology Application Center (MITAC)

P.O. Box 980480
Richmond, VA 23298
(804) 827-1020
http://www.meditac.com
Focus: Develops and evaluates telemedicine technology that will benefit the practice of medicine in space and on Earth.

ProVision Technologies (PVT)

Bldg. 9313 Room 130
Stennis Space Center, MS 39529
(228) 689-8176
http://www.pvtech.org
Focus areas: Applies hyperspectral imaging technology as a noninvasive diagnostic tool to aid in the following: human stress detection, wound healing, skin cancer treatments, food contamination detection.

Solidification Design Center (SDC)

Professor of Mechanical Engineering
Auburn University
201 Ross Hall
Auburn, AL 36849
(334) 844-5940
http://metalcasting.auburn.edu/
Focus areas: Improves the casting of metals and alloys by researching everything from the metal proportions to the solidification process.

Space Communications Technology Center

Florida Atlantic University
Boca Raton, FL 33431
(561) 297-2343
Web site: http://www.fau.edu/divdept/comtech/ctchome.html
Focus: Develops commercial digital transmission techniques for video, audio, and data from satellites to Earth.

Space Vacuum Epitaxy Center (SVEC)

Space Vacuum Epitaxy Center
University of Houston
Science and Research Building One
4800 Calhoun Road
Houston, TX 77204-5507
(713) 743-3625
http://www.svec.uh.edu/
Focus: Uses the natural vacuum of space to research and develop the following: semiconductor mid-IR lasers, high-efficiency solar cells, oxide thin films, ultra-hard thin film coatings.

Wisconsin Center for Space Automation and Robotics (WCSAR)

Wisconsin Center for Space Automation and Robotics
College of Engineering, UW-Madison
1500 Johnson Drive
Madison, WI 53706
(608) 262-5526
http://wcsar.engr.wisc.edu/
Focus: Supports agribusiness research and provides industry partners with the following: advanced environmental control technologies, large-scale controlled environment plant production facilities, automation technologies.

Appendix E

Lesson Plans

The following are some examples of the lesson plans NASA provides to teachers and students through its NASAexplores Web site. The following lesson plans are for grades 9–12; NASAexplores also has lesson plans on the same topics for younger students. Lesson plans include an introductory article written specifically for the lesson, teacher sheets, and student sheets. Each lesson cites the applicable science, math, technology, or geography standards met in the lesson. Those National Education Standards can be found following the final lesson plan.

LESSON 1

The Air That We Breathe

Courtesy of NASA's Aerospace Technology Enterprise

What's the concern about carbon dioxide? It's a colorless, odorless, tasteless, non-toxic gas that we exhale as we breathe. Plants use it for photosynthesis. It's an important element of the Earth's climate. Carbon dioxide (CO_2) is also produced by combustion: burning fuels produces CO_2, for example. Industry and transportation create great amounts of CO_2. Heating our homes, operating vehicles, and cutting away large expanses of vegetation (no plants means no photosynthesis) all add to the amount of CO_2 in our atmosphere.

Carbon dioxide is called a greenhouse gas because it's one of the gases in our atmosphere that helps keep the Earth warm—much the same way a greenhouse keeps plants warm. When the Sun's energy penetrates our atmosphere, some of the energy is reflected back while some is absorbed by the Earth's surface. Some of these energies remain trapped in our atmosphere, warming the planet. This is called the greenhouse effect, and without it, humans couldn't survive on Earth.

CO_2 becomes a problem, however, when more CO_2 is produced than used. The extra carbon dioxide and other gases that we release into the atmosphere help increase the amount of energy that becomes trapped. When too much warmth is held in the atmosphere, the Earth's temperature rises and that isn't good. The enhanced greenhouse effect is also referred to as global warming.

How Is NASA Helping Reduce CO_2 Emissions?

NASA can't reverse the Industrial Revolution, and it can't stop population growth, but the agency can work to design aircraft engines that operate more efficiently and produce fewer CO_2 emissions. At NASA's Glenn Research Center in Ohio, the Ultra-Efficient Engine Technology (UEET) Program's goal is to reduce aircraft CO_2 emissions by 8 to 15 percent.

To do this, engineers are making jet engine compressors, turbines, and, nozzles more efficient by reducing weight. The less a jet weighs, the less fuel it will require to fly. One weight-saving approach is the development of new compressor designs, using advanced, computer-generated analytical techniques. UEET engineers are researching ways to add more load to the compressor blades to make the compressor shorter, requiring fewer stages to do the same amount of work. The shorter the compressor is, the less the engine weighs. Another weight-saving measure involves the development of advanced lightweight metallic structures and new ceramics. Replacing engine materials with these advanced materials will lighten the engine.

UEET materials researchers are also developing advanced high-temperature materials to allow an engine to operate at higher temperatures, making the engine more efficient and burn less fuel. Burning less fuel reduces CO_2 emissions.

UEET engineers are reducing aerodynamic drag—the resistance force that makes aircraft use more energy to fly—by designing components that make better use of fuel. Lowering drag improves air vehicle performance and efficiency, which reduces the amount of fuel burned. That translates into better fuel mileage and reduced CO_2 emissions.

NASA hopes to have the designs tested and operational in jets by the year 2020.

Lesson: Greenhouse Effect

Objective: To explore how global warming occurs and its impact on Earth.

Grade Level: 9–12

Subject(s): Earth Science, Chemistry, Technology

Prep Time: $<$ 10 minutes

Duration: 1 week

Materials Category: Special

National Education Standards:

Science: 3c, 7d, 7e, 7f

Technology (ISTE): 6, 10, 14

Materials:

- Research materials, such as Internet access, science books, and magazines
- Student Sheets

Related Link(s):

The Earth: A Giant Greenhouse
http://seawifs.gsfc.nasa.gov/SEAWIFS/LIVING_OCEAN/TEACHER5.html

Student Sheet(s)

Global warming is the gradual increase in temperatures on Earth. Generally, this warming is attributed to the increase of greenhouse gases in the Earth's upper atmosphere. Some greenhouse gases occur naturally in the atmosphere, while others result from human activities. Naturally occurring greenhouse gases include water vapor, carbon dioxide, methane, nitrous oxide, and ozone. Certain human activities, however, add to the levels of most of these naturally occurring gases.

However, some solar scientists are considering whether the warming exists at all. And, if it does, it might be caused, wholly or in part, by a periodic but small increase in the Sun's energy output. An increase of just 0.2% in the solar output could have the same effect as doubling the carbon dioxide in the Earth's atmosphere.

In order to help you form your opinion, answer these questions by conducting research on global warming.

1. Draw a diagram that explains the causes of global warming.
2. The greenhouse effect is natural, why is it a problem?
3. What is the evidence that global warming exists? How reliable and accurate is this evidence?
4. What are the projected long-term effects of global warming?
5. What is the evidence that global climate change might be affected by solar variation?

6. What can or should be done about global warming if it exists and is caused by pollutants and emissions? (Even if global warming does not exist, are there other reasons for lessening the emission of pollutants into our atmosphere?)

7. Catalog the greenhouse gases. List their chemical formulas, molecular structures, sources, relative abundance, stability, and overall impact to global warming.

8. List the major contributors of carbon dioxide (natural, human induced). Rank countries according to their contribution to the problem.

9. In some ways, the issue of global warming is controversial. Explain the controversy by exploring outstanding questions that remain. Summarize the "knowns" and "unknowns."

10. How has the issue affected public policy? What agencies have formed? What conferences have been held? Is the issue becoming political?

Teacher Sheets

Pre-lesson Instructions

Reserve the school library or computer room for two class periods for students. Ensure the computers have Internet access.

Guidelines

1. Read the 9–12 NASAexplores article, "The Air We Breathe." Ask students to write a one to two paragraph summary of the article.

2. Hand out the Student Sheets. Explain to students that they will be researching the greenhouse effect.

3. Provide students with time and resources to find the answers to the questions.

4. After an appropriate amount of research time, have students turn in their work.

Discussion/Wrap-Up

1. Ask students to share what they learned while answering the questions.

2. Have the class vote on whether they think global warming is due to increased amounts of greenhouse gases or an increase in solar output. Ask students to explain the reasoning behind their answer.

3. As a class, make a list of regulations and changes that have occurred since the concept of global warming became public interest.

4. Review how NASA is addressing the issue.

Extensions

- Complete the 9–12 NASAexplores lesson, "Detection of Carbon Dioxide."
- Build a greenhouse.
- Make a concept map using the following terms: greenhouse effect, temperature, gases, atmosphere, Earth, space, infrared rays, methane, carbon dioxide, nitrous oxide, ozone, water vapor, etc.

Glossary

carbonate—a salt or ester of carbonic acid.

drag—resistance to motion through the air.

fermentation—the anaerobic conversion of sugar to carbon dioxide and alcohol by yeast.

infrared—the range of invisible radiation wavelengths from about 750 nanometers, just longer than red in the visible spectrum, to 1 millimeter, on the border of the microwave region.

LESSON 2

Healing Light

Courtesy of NASA's Human Exploration and Development of Space Enterprise

Experiments with light energy are making everyone happy: farmers, doctors, patients, and astronauts are all reaping the benefits. What began as a way to grow better crops has ended up being a way to help seriously ill patients recover faster. In this case, what's good for plants is also good for people.

It started with Light Emitting Diodes (LEDs) developed by NASA Marshall Space Flight Center in Alabama and Quantum Devices, Inc., of Wisconsin. The scientists exposed plants aboard the Space Shuttle to the near-infrared light produced by LEDs. They found that the LEDs increased the energy produced in the mitochondria (energy compartments) of each cell. That meant the cells grew faster. Faster-growing plants are good news for farmers; the faster the plants grow, the sooner they can be harvested, processed, and sold.

Right about the same time, Quantum Devices scientists heard physicians discussing the use of laser therapy for their patients. While laser light did accelerate cell growth and healing in patients, there were some significant drawbacks. Lasers can cause tissues surrounding the treatment area to become overheated, they're big and expensive, limited in wavelength (color), and they're not very reliable, said Harry T. Whelan, MD, professor of pediatric neurology and director of hyperbaric medicine at the Medical College of Wisconsin. A light went off, so to speak, and Quantum approached Dr. Whelan about his concerns. Soon they were experimenting to see if using LED instead of laser therapy would improve the quality of treatment for patients.

"LED treatment has been a wonderful advancement," Dr. Whelan says. "LEDs don't heat the tissues the way lasers do; because LED uses longer wavelength (redder) near-infrared light, it penetrates the tissues deeper. And where lasers are more pinpointed in their delivery, LED can treat the entire body. That's useful for treating people with serious burns, crush injuries, and complications of cancer chemotherapy and radiation treatment, where large portions of the body are involved."

LED therapy has been used successfully with diabetic skin ulcers, burns, and severe oral sores caused by cancer treatment. The redder the light, the longer the wavelength, and the longer the wavelength, the more deeply it can penetrate body tissues, Dr. Whelan says. The near-infrared light rays produced by LED are longer than (and therefore superior to) lasers, and Dr. Whelan asserts that this improved therapy could extend to treating brain tumors and injuries. Animal experiments being conducted now direct LED through the head without the use of any surgery. When LED light is used to activate light-sensitive chemotherapy drugs to destroy cancer, it is dubbed Photo Dynamic Therapy (PDT). LED light is otherwise used without drugs to stimulate normal cell chemicals for healing and tissue regeneration.

"LED reacts with cytochromes in the body," says Dr. Whelan. "Cytochromes are the parts of cells that respond to light and color. When cytochromes are activated, their energy levels go up and that stimulates tissue growth and regeneration. The potential to regenerate tissue, muscle, brain, and bone opens the door to helping people with diseases that previously had no hope of treatment."

The good news about using LED therapy to speed healing made its way back to the space program. Muscle and bone atrophy are well documented in astronauts because microgravity slows the healing process and alters the function and structure of every cell's mitochondria, Dr. Whelan says. The result is that wounds are slow to heal, and muscles and bones become weaker from time spent in space. The idea of using LED therapy with astronauts sounded appealing.

"Using an LED array to cover an astronaut may help prevent the effects of microgravity," says Dr. Whelan. "LED therapy could also be used to help treat conditions that could arise in space that don't respond to treatment because of those microgravity situations. A simple cut might heal faster with LED, but the benefits would be even more notable if an astronaut suffered a severe injury."

Here on Earth, Dr. Whelan says that LED therapy can easily affect our entire population. "Not everyone may need to use LED treatments for themselves, but just about everyone has known someone with cancer or a severe injury," he says. "Knowing that there is hope for diseases that used to have no treatment is good news for everyone."

Lesson: Light Effect on Plant Growth

Objective: To observe plant growth using different colored wavelengths of light.

Grade Level: 9–12

Subject: Biology

Prep Time: 10–30 minutes

Activity Duration: 3 weeks

Materials Category: Special requirements

National Education Standards: Science 1b, 4e

Materials:

- Radish seeds
- Potting soil
- Large styrofoam cups or cans
- Metric ruler
- Graph paper
- Small boxes
- Red, green, blue and clear transparent plastic wrap

Student Sheet(s)

Background

Light controls many plant processes. For example, light in the 400–450 nm (blue) and 625–700 nm (red) wavelengths is required for photosynthesis. Photoperiod responses occur in the red wavelengths. Phototropic (growth of plants toward the light source) responses occur in the blue wavelengths. Plants are grouped into three areas based on photoperiod responses; short-day, long-day, and day-neutral. Short-day plants initiate flower buds when days are shorter. Long-day plants continue in a vegetative state until day length is longer. Flowering for day-neutral plants is not controlled by photoperiod.

When given light for 24 hours, short-day plants continue to grow vegetatively, and flowering is inhibited. Long-day plants are induced to flower earlier, and continue to put on vegetative growth. Day-neutral plants continue to grow vegetatively. A concern with landscape lighting is that plants will continue vegetative growth late in the season. This late growth will not have the opportunity to prepare properly before winter, and may experience winter injury.

Procedure

1. Punch six holes around each cup or can one inch from the top, so that air is able to circulate.

2. Each group will prepare four pots with potting soil.

3. Students will plant 10 seeds just below the surface of the soil in each pot. Water equally.

4. Place containers near a window receiving full sunlight.

5. Every two days, gently pour 1/4 cup of water into each container.

6. After seeds begin to sprout, cover the containers with cellophane. Allow at least 3–5 days for germination and initial development.

7. Place the four pots next to each other. Select 5 plants, all close to the same size, from each pot by weeding out the rest.

8. Label each pot: A, B, C, and D.

9. Measure the height of each plant in centimeters and record the average height for each pot in the chart below.

10. Place 1A in a box and cover it with red clear wrap.

11. Place 1B in a box and cover it with green clear wrap.

12. Place 1C in a box and cover it with blue transparent wrap.

13. Place 1D in a box and cover it with clear wrap.

14. Place the boxes in the sunlight and measure the growth of the plants every other day for 21 days in millimeters. Record your data in the chart below.

15. Graph the data after the 21-day trial period.

Day	Pot A	Pot B	Pot C	Pot D
1	______	______	______	______
3	______	______	______	______
5	______	______	______	______
7	______	______	______	______
9	______	______	______	______
11	______	______	______	______
13	______	______	______	______
15	______	______	______	______
17	______	______	______	______
19	______	______	______	______
21	______	______	______	______

Questions

1. Which plant group showed the fastest growth rate?
2. Which group showed the slowest?
3. What was the control in the experiment?
4. If you wanted to increase crop yield what light exposure would be the most help?

Teacher Sheet

Additional Background

Incandescent lights produce all wavelengths of light, and are closest to natural sunlight. Incandescent is the least expensive light source to purchase, but the most expensive to operate because of short lamp life and low efficiency. Interestingly, it still remains the most popular type of lighting for residential landscapes. Metal halide is more efficient to operate and has wavelengths in the blue-green-red area. It has better color emission than mercury, but not quite as good as incandescent. Mercury vapor lamps emit light in the blue-green range, which accentuates the green color of plants.

Some things can be done to avoid possible problems with landscape lighting. Select the proper light for the area. For security purposes, high-pressure sodium lights are the best choice. Metal halide would be preferred over incandescent in residential areas, malls, parks, etc., where true color is important. Lights that may affect plant growth can be shielded to direct light away from plants. Many plants, long-day and day-neutral species, are better adapted to security lighting locations.

Typical Results

The tallest plants are covered in blue cellophane. The second tallest are covered in red cellophane. The full-sunlight plants rank third in height. The green covered plants are the shortest. The plants with the most foliage are the ones exposed to full light. Again the green covered plants had the least amount of foliage.

Extension

Try the same experiment using short-day, long-day, and day-neutral plants. Experiment with different light sources.

Glossary

chemotherapy—treatment of abnormal cells and invading organisms by chemical compounds.

cytochrome—any of several intracellular respiratory pigments that are enzymes functioning in electron transport as carriers of electrons, which respond to light and color.

diodes—semiconductor device.

hyperbaric—an agitated bodily condition when the pressure within the body tissues, fluids, and cavities is countered by a greater external pressure, such as may happen in a sudden fall from a high altitude.

mitochondria—mobile cytoplasmic organelles of eukaryotes visible in the light microscope whose main function is the generation of ATP molecules.

pediatric neurology—the study of the nervous system in children.

LESSON 3

Moving on Down the Highway

Courtesy of NASA's Aerospace Technology Enterprise

The next time you're riding down an interstate highway and notice a big rig, take a closer look. Most of the more modern trucks have aerodynamic shielding, called fairings, over the cab of the vehicle to help air flow more efficiently, reduce drag, and increase fuel mileage. Those improvements came about in part because of NASA research.

Research done at NASA in the 1970s influenced the newer shapes of heavy-duty vehicles. Developing aerodynamically superior designs provided fuel savings—20 to 24 percent—without having to change the cargo or the interior structure. By the late 1990s, the clean lines in trucks gained acceptance in the trucking industry, and interestingly, the original NASA designs were also proposed for the design of better livestock hauling vehicles.

What makes a truck more aerodynamic? Metal, fiberglass, or plastic fairings round the front corners, and place smooth edges on the cab's roof and sides extending back to the trailer. These fairings minimize the space between a truck's cab and the trailer. By minimizing gaps and sharp corners, air flows much more smoothly to the back of the vehicle. Older trucks had protruding edges and sharp rises in contours that blocked airflow and reduced fuel mileage. These changes resulted in 20 to 24 percent lower fuel consumption than standard vehicles, yet did not significantly affect the interior load-carrying capacity of the trailer.

Further research showed that extending the rear of a vehicle slightly, so the end gradually tapered, further reduced drag. This "boat tail" structure brought the potential for fuel savings up to 27 percent.

During testing, engineers created models of an ideal truck design that took advantage of aerodynamics in the most efficient way. It wasn't a functional design for real-world use, but it illustrated the design changes that could make a difference in truck performance. From there, they came up with a plan for what would really work with trucks on the highway.

Various parts of the truck assembly were evaluated in wind tunnels. Engineers noted that any place that trapped or diverted air was a likely drain on fuel efficiency. Most of the time, NASA technology used in the civilian or commercial world is first used in space or aerospace projects, and is then "spun off" or transferred to the commercial sector. For the fuel economy project for road vehicles, the technology transfer was not as straightforward.

In 1973, America—and the world—experienced an energy shortage from an oil embargo and rapidly rising fuel prices. There was a general call to develop more fuel-efficient designs, and because of NASA's experience in creating aerodynamic designs for high-speed jets, the space agency tackled the down-to-Earth issue of reducing highway-driving costs.

It took over twenty years for these changes to the truck fore body to be implemented in over-the-road trucking. Re-designing that many large vehicles is an expensive proposition, and was introduced slowly, to save costs. But, the eventual cost savings is enormous. In one year's time, if all van or box-shaped trucks in the United States used this setup, it would save 26 million barrels of fuel. That would allow over 1.3 million cars (assuming they achieved 30 miles per gallon) to drive 25,000 miles—the distance around the world.

Lesson: Fairings

Objective: To examine an aerodynamic feature of vehicles.

Grade Level: 9–12

Subject(s): Physical Science, Physics, Technology

Prep Time: < 10 minutes

Duration: 1 week

Materials Category: Special

National Education Standards:

Science: 3d, 6a, 6b

Technology (ITEA): 3a, 3b, 5b

Materials:

- Camera
- Film

Related Link(s):

AeroTruck Photo Gallery Contact Sheet
http://www.dfrc.nasa.gov/gallery/photo/AeroTruck/HTML/index.html

Student Sheet(s)

Background

Large trucks consume lots of fuel in overcoming aerodynamic drag and rolling resistance. As the truck goes faster, the horsepower needed to overcome aerodynamic drag becomes paramount, increasing with the cube of velocity. Fuel consumption is proportional to required power, so efforts to overcome drag are important for boosting fuel economy. Previous efforts have mainly involved the use of fairings. Fairings are secondary structures that are added to reduce drag by streamlining a vehicle. To learn more about fairings, read the article, "*Moving on Down the Highway.*"

Questions

1. What is drag?

2. How would drag impact an automobile or truck?

3. What is a fairing?

4. How does a fairing reduce drag?

Procedure

1. Each group of two to three students will use a camera to take five pictures of large trucks that exhibit fairings.

2. Under each picture, explain how the fairing increases the aerodynamics of the truck.

3. Take one additional photograph of a vehicle you feel exhibits a poor aerodynamic design. Explain under the picture, why you feel this vehicle is an example of a poor aerodynamic design.

4. For a conclusion, write two paragraphs explaining the impact fairings have made on the trucking industry.

5. If your group is using a digital camera, the photos may be incorporated into a PowerPoint presentation. Your work may also be presented on a poster board or in a book format.

Teacher Sheets

Background

During the NASA investigation of truck aerodynamics, the techniques used in flight research proved highly applicable. By closing the gap between the cab and the trailer, researchers discovered a significant reduction in aerodynamic drag, one resulting in 20 to 25 percent less fuel consumption than the standard design. Many truck manufacturers subsequently incorporated similar modifications to their products.

Frontal pressure and a vacuum forming in the rear of the vehicle cause drag. Frontal pressure is caused by the air attempting to flow around the front of the vehicle. As millions of air molecules approach the front grill of the vehicle, they begin to compress. This raises the air pressure in front of the vehicle. At the same time, the air molecules traveling along the sides of the vehicle are at a lower pressure compared to the molecules at the front. The compressed molecules will move from a high-pressure zone to a low-pressure zone. To achieve this, the molecules move from the front to the sides, and the top and bottom of the vehicle. A vacuum in the rear of the vehicle is also present. This is caused by the "hole" left in the air as the vehicle passes through it. A forward-moving vehicle punches a big hole in the air. This causes air molecules to rush around the body of the vehicle.

The space directly behind the vehicle is "empty" or like a vacuum. This empty area is a result of the air molecules not being able to fill the hole as quickly as the vehicle can make it.

Guidelines

1. Read the 9–12 NASAexplores article, "Moving on Down the Highway" with students. Discuss how NASA technology has been used to improve the trucking industry.

2. Hand out the Student Sheets, and divide students into groups of two to three.

3. Have students brainstorm on places they can take pictures of commercial trucks.

Discussion/Wrap-Up

Have students present their work to the class.

Extensions

- Examine aerodynamic features added to race cars and bicycles.
- Interview a truck driver. Ask the driver questions on gas mileage and the designs of trucks.

Glossary

contour—the outline of a figure, body, or mass.

fairing—an auxiliary structure on the external surface of a vehicle, such as an aircraft, that serves to reduce drag.

LESSON 4

Quieting the Roar

Courtesy of NASA's Aerospace Technology Enterprise

Here's a practical tip: never hold a conversation while a jet is flying overhead. Why? You'll have a hard time hearing what the other person is saying because the jet engines are so loud. If NASA projects go according to plan, that may soon change. A group of programs called Quiet Aircraft Technology (QAT) has the 5-year goal of reducing aircraft noise by 50 percent.

Led by NASA's Langley Research Center in Virginia, key participants in the QAT program include Glenn Research Center in Ohio and Ames Research Center in California. NASA Glenn has the primary responsibility for the QAT Engine System Noise Reduction Project. As you might suspect, Glenn engineers are focusing on modifying the jet engines to reduce aircraft noise. Program manager Joe Grady thinks they'll succeed in meeting the goal.

"An engine is a large machine with two sources of noise," he says. "When a jet flies towards you, the fan pulling air into the engine creates the characteristic high-pitched whining sound we are familiar with. When the airplane flies away from you, that lower-pitched rumble is created by the jet exhaust exiting from the engine." With multiple sources of noise, it stands to reason that there would be many approaches to solving the issue—and there are.

One of the most promising and revolutionary ideas under development at NASA to reduce fan noise is called trailing edge blowing. A primary source of fan noise is the interaction of the aerodynamic wake created by the fan blades with other engine components downstream. Researchers have found that the noise created by these interactions can be reduced significantly by blowing air through the fan's trailing edge to reduce the strength of the pressure gradient, or wake, behind the fan blades. Preliminary evaluations in NASA test rigs confirm the potential of this approach achieve a breakthrough in reduction of fan noise. Jet noise is caused by turbulence in the exhaust, exiting the engine at high speed and temperatures. Mixing of the engine exhaust with the surrounding air reduces the turbulence and decreases the resulting noise. How do they achieve this? One approach that shows promise is nozzle chevrons. The addition of the saw tooth chevron shape to the exhaust nozzle exit causes the hot engine air to mix more thoroughly with the cooler ambient air. The turbulence of the exiting air is decreased, and the result is a reduced level of jet noise. Recent flight tests have shown that this simple solution reduces noise significantly.

The focus of NASA's Quiet Aircraft Technology program is to develop and demonstrate the promising noise reduction concepts in tests. Glenn Research Center facilities include wind tunnels and acoustic chambers to conduct the initial evaluations of noise reduction technologies, and concepts that show promise will then be evaluated in engine tests. Flight test evaluations are conducted before a new concept or technology is actually incorporated into a product engine.

As a result of the current research, NASA expects to have several engine noise reduction technologies and concepts demonstrated by 2005. "The challenge will be to take successful NASA technology and integrate it into real products used by the aircraft industry," Grady says. "Technology helps no one if it's not used in the real world."

Lesson: Hear Me

Objective: To describe the anatomy and function of the human ear, and to outline the path of a sound wave through the ear.

Grade Level: 9–12

Subject(s): Biology, Life Science

Prep Time: < 10 minutes

Activity Duration: Two 50-minute class periods

Materials Category: General classroom

National Education Standards: Science: 4e, 4f

Materials:

- Biology textbook
- Other resources and Internet access (optional)

Related Link(s):

Department of Otolaryngology-Head and Neck Surgery at the University of Washington School of Medicine: http://depts.washington.edu/otoweb/ear_anatomy.html

The Soundry: http://library.thinkquest.org/19537/Main.html

Student Sheet(s)

Background

The impact of noise on hearing, health and the quality of life can no longer be disputed. Volumes of literature exist to show the hazards to hearing from repeated exposure to noise. One in ten Americans has a hearing loss that affects their ability to understand normal speech. Excessive noise exposure is the most common cause of hearing loss. The damage caused by noise is called sensorineural hearing loss, or nerve deafness.

In order to understand how noise can damage the ear, you must understand how the ear functions.

Procedure

1. What is the scientific name of the sense of hearing (e.g., smell = olfactory)?

2. Using your textbook and other resources, create a diagram of the organ used for hearing, and include the names of the different parts of this organ.

   Describe the functions of the following:

   - Pinna
   - Tympanic membrane
   - Ear ossicles
   - Oval window
   - Cochlea
   - Organ of Corti
   - Auditory nerve

3. Outline the path of a sound wave through the external, middle, and inner ear, and identify the energy transformations that occur. This can be done by creating a flow chart (using words and arrows) indicating how this organ receives and understands the sensory information as it travels from outside a person through the organ.

4. Explain how the brain interprets the messages received by this organ and how it helps interpret these messages.

Internet Resources

Department of Otolaryngology-Head and Neck Surgery at the University of Washington School of Medicine (http://depts.washington.edu/otoweb/ear_anatomy.html)

The Soundry (http://library.thinkquest.org/19537/Main.html)

Teacher Sheets

Pre-lesson Instructions

1. As students enter class, have a tape recorder playing of the sound of a cricket. Hide the tape player. Start class as normal. As students begin to take notice or comment on the cricket sound, ask students, "Does this sound bother you? If so, why? Where is the sound coming from?"

2. Ask students to define the term "noise." "Would the cricket sound be considered noise?"

3. Explain to students that what one person considers noise another may not. (For example, rap or heavy metal music.) Discuss how loud noises can damage a person's hearing.

Background

A source of sound sends sound waves into the air. The sound waves travel through the ear opening, down the ear canal, and strike the eardrum, causing it to vibrate. The vibrations are passed to the small bones of the middle ear, which transmit them to the hearing nerve in the inner ear. Here, the vibrations become nerve impulses and go directly to the brain, which interprets the impulses as sound.

When noise is too loud, it begins to kill the nerve endings in the inner ear. As the exposure time to loud noise increases, more and more nerve endings are destroyed. As the number of nerve endings decreases, so does a person's hearing. There is no way to restore life to dead nerve endings; the damage is permanent.

Guidelines

1. Read the NASAexplores article, "Quieting the Roar." Discuss why NASA would be interested in reducing the noise level of airplanes. Why is the sound from an airplane considered to be noise pollution?
2. Hand out the Student Sheets.

Discussion/Wrap-Up

- Have students orally summarize their findings.
- Discuss other sources of noise pollution and measures that have been taken to reduce the noise level. For example: sound barriers along highways reduce noise levels in surrounding neighborhoods, effectively halving the loudness of vehicles. Noise barriers line approximately 1400 miles of U.S. Federal highways alone, with the number growing by 90–100 miles every year.

Extension(s)

- Make a list of ways to reduce the noise level on the school campus. Organize free hearing screenings.
- Contact local hospitals, speech and hearing clinics, or universities for speakers.

Glossary

thrust—the forward-directed force developed in a jet or rocket engine as a reaction to the high-velocity rearward ejection of exhaust gases.

wake—a track, course, or condition left behind something that has passed.

LESSON 5

Building Better Golf Balls

Courtesy of NASA's Aerospace Technology Enterprise

Sometimes space technology brings remarkable discoveries that make the world a better place, from learning more about ozone depletion to developing medical technologies to save lives.

Other times space-related innovations just add to the fun factor of our lives. Two NASA technology applications have done just that: because of them, it is now possible to build better golf balls. The Glenn Imaging Technology Center, part of NASA Glenn Research Center in Ohio, developed high-speed video equipment to measure, analyze and obtain accurate data that was originally used in both the lunar missions and in aircraft engine development.

Instead of filming the aircraft engines at normal speeds, for instance, cameras recording thousands of images per minute allowed scientists to see smaller details and track that information more specifically.

When NASA develops technology that could be used in the commercial world, it advertised through the Technology Transfer program. Possible applications, social and economic benefits and a summary of the science involved are highlighted, and interested organizations are invited to learn more about how space and aerospace research can benefit people on Earth. NASA developers showcased their high-speed imaging technology through the Technology Transfer program and the result was an inquiry from a sporting manufacturer.

The Ben Hogan Company's Golf Ball Division had already developed several new experimental ball designs, but was unable to determine exact performance characteristics. NASA's high speed video equipment, advanced computer hardware, and software provided the perfect tools to measure spin rates of the seven experimental golf balls, as each was hit by four different golf clubs. The spin rate calculations helped Ben Hogan evaluate the designs.

The golf balls were marked with analysis control points, and measured as they were hit several times with different clubs that simulated the striking patterns of novices, duffers and professionals. The high-speed video equipment captured images of balls in flight, recorded and analyzed the data, and presented spin rate and velocity information that allowed Ben Hogan to improve the spin characteristics of the ball. The final reports showed how textured patterns in the ball's surface reduced friction and affected the aerodynamics, and how variations in the liquid center of the ball affected flight.

In another golf-related venture, Wilson Sporting Goods' goal was to create the most symmetrical ball surface available, which they felt would lead to increased distance and accuracy. Using technology similar to that used to test the Space Shuttle's external tanks, Wilson engineers researched the size, depth, and shape of the dimples, which are the bumpy patterns on the surface of a golf ball. Three-dimensional computer graphic simulation examined the aerodynamics of the golf balls, and the engineers found that large dimples reduced air drag, enhanced lift, and maintained spin for distance; small dimples prevented excessive lift that would destabilize ball flight; and medium-sized dimples blended the characteristics of small and large dimples. This information led Wilson designers to significantly increase the number of dimples per ball and to include a variety of sized dimples on each ball, distributed strategically over the surface.

The results of all this high-tech testing? Golf balls that go faster and farther, which means lower scores and a more enjoyable game. And you thought the only connection between astronauts and golfing was that famous moon-shot taken by Alan Shepard on the *Apollo 14* mission.

Lesson: Magnus Lift

Objective: To understand the concept of Magnus Lift and to be able to explain how this concept affects several sport games.

Grade Level: 9–12

Subjects: Physical Science, Physics, Technology

Prep Time: <10 minutes

Activity Duration: 40 minutes

Materials Category: Common classroom

National Education Standards: Science 1e, 3d, 6a; Technology/ITEA 3a, 3b, 8a, 9a, 9c, 10a

Materials:

- Low mass ball, such as a Styrofoam ball
- Same size ball but with a heavier mass
- Paper & pencil

Student Sheet(s)

Background

Magnus Lift is a force experienced by a spinning ball or cylinder in a fluid. The effect is responsible for the swerving of golf and tennis balls when hit with a slice. The Magnus Lift also explains why curve balls curve.

Magnus Lift is similar to Bernoulli's Principle. Bernoulli's Principle states that fluid pressure decreases at points where the speed of the fluid increases. For a spinning baseball, the stitches on the ball, cause pressure on one side to be less than on its opposite side. This forces the ball to move faster on one side than the other and the ball will curve.

A German physicist and chemist H.G. Magnus experimentally investigated the effect. In 1853, Magnus reported that when a spinning object moves through a fluid (such as air), it experiences a sideways force.

A spinning object moving through a fluid departs from its straight path because of pressure differences that develop in the fluid as a result of velocity changes induced by the spinning body. In the case of a ball spinning through the air, the turning ball drags some of the air around with it. Viewed from the position of the ball, the air is rushing by on all sides. The drag of the side of the ball turning into the air (into the direction the ball is traveling) retards the airflow; on the other side, the drag speeds up the airflow. Greater pressure on the side where the airflow is slowed forces the ball in the direction of the low-pressure region on the opposite side, where a relative increase in airflow occurs.

The spin of a golf ball alters the shape of the flow of air around the ball. As a result of spin, the topside of the golf ball is moving away from the direction of the airflow, and the bottom side is moving toward the direction of the airflow. As a result, the relative speed of the air on the top is much greater than the relative speed of the air on the bottom. A low pressure is created on top of the golf ball, and a higher-pressure region is created below the ball. The net force resulting from this is known as aerodynamic lift. To summarize the Magnus Lift is the generation of a lifting force perpendicular to the axis of rotation of a spinning object.

The dimples increase drag of the golf ball, thereby increasing the Magnus Lift. Without them, the ball would travel in a more parabolic trajectory than an impetus trajectory, hit the ground sooner, and not drop straight down.

Procedure

1. Throw the ball forward with as much spin as possible.
2. Try to make the ball curve in different directions by adjusting its axis of rotation.
3. Try steps one and two again with a heavier ball.

Questions

1. Explain the forces acting on the light and heavy balls when thrown.
2. How were the forces different for the different mass balls?
3. Draw a picture of a golf ball showing the air pressure around the ball.
4. Summarize the Magnus Lift. How does it impact sports such as golf and baseball?

True/False Quiz

1. Regardless of how you "tilt" a ball it still looks the same—you can't change the angle of attack on a spherical ball.

2. To generate lift on a ball, it is necessary to set up an imbalance. Spinning the ball will do this.

3. If a ball wasn't rotating as it flew through the air, then both the top and bottom sides of the ball would meet the air rushing over it at the same speed.

4. For a curve ball, the top of the ball spins forward into the oncoming air.

5. For a curve ball, there is less movement of the air toward the bottom surface.

6. With high pressure on one side and low pressure on the other, there is an imbalance in the forces on the ball.

7. Magnus Lift is when the higher pressure on the top curves the ball downward from its straight-line path.

8. The Magnus Lift affects only some sports balls.

9. The Magnus Lift will pull a ball down faster.

10. If there is no spin, there is no lift.

Teacher Sheets

Guidelines

1. Read the article, "Building Better Golf Balls," then discuss the article with students.

2. Show a short video clip of a baseball pitcher throwing a ball or of a golf player hitting the ball. An alternative would be to have a baseball or golf player demonstrate the act for the students. Discuss the motion of the ball and the motion of the player. Ask students,"What forces are acting on the ball?" "How can the direction of the ball be changed?"

3. Find a large area outside or in the gym for students to complete the procedural part of the lesson.

4. After students have thrown the ball and had a chance to answer the questions, explain the dynamics of curve balls. When the students throw the ball it should curve in flight toward the side that's heading back toward the student as the ball spins. As the ball spins, it experiences the Magnus Lift and a wake deflection force. These forces will together push the ball toward the side that is heading back to the pitcher. A ball with a lower mass will accelerate easily and curve substantially. A heavier ball will not curve as dramatically.

Answers to True/False Questions

1. T
2. T
3. T
4. T
5. F
6. T
7. T
8. F
9. T
10. T

Extension

Invite a local high school or college baseball or softball pitcher to demonstrate different pitches and explain how the proper spin is achieved.

Related Links

Golf Aerodynamics: http://www.jsc.nasa.gov/er/seh/pg76s95.html

Magnus Effect—Flettner's Ship: http://www.physics.umd.edu/deptinfo/facilities/lecdem/f5-31.htm

Cislunar Aerospace: http://wings.ucdavis.edu/Tennis/Book/magnus-01.html

LESSON 6

Wanted: Green Airplanes

Courtesy of NASA Explores

NASA is working to create green airplanes—green as in environmentally friendly, that is. One of the biggest environmental issues with jet airliners is the Nitrogen Oxide Emissions (NOx for short) spewed into the atmosphere during takeoff and landing procedures.

"While we develop technology for supersonic and hypersonic aircraft engines our major emphasis for study right now is the subsonic aircraft engine," says Lori Manthey, program support officer for the Ultra-Efficient Engine Technology (UEET) Program Office at Glenn Research Center in Ohio. "Subsonic planes affect local air quality with NOx emissions, and that's shown in smog and human health issues. You can see smoke coming from older airplanes, but for the most part, NOx emissions are invisible in modern clean-burning engines. The pollution is still there, though, and we want to reduce the damage it causes to local air quality."

Airports and environmental agencies measure NOx emissions, Manthey says. Polluting airlines are charged a penalty when they use a stricter country's airport, so there is a financial incentive as well as an ethical reason to be concerned with air quality. "The goal is to reduce NOx emissions during landing and takeoff to 70 percent below the international standards created in 1996," Manthey says. "We're not there yet, but we're making very good progress."

Computer models of jet engines help scientists and engineers understand the variables. From there, engineers use flame tubes to duplicate the way fuel is injected into a jet engine's combustor. "Compressed air is ignited with fuel, and that is what provides the thrust for an aircraft," she says. "Our experiments with flame tubes have shown that we can reduce NOx emissions below that 70 percent mark by using advanced combustor designs in engines. That's encouraging news, and now that we've found this in flame tubes, we'll go on to the next step."

The NOx emissions put out by aircraft are minor when compared to land vehicles, Manthey says. Cars, trucks, buses and trains contribute far more to air pollution, primarily because there are so many more land vehicles than aircraft. "But we expect air traffic to double and triple by 2010," she says, "so the number of airplanes and flights will grow tremendously. That means that NOx emissions from aircraft will become a larger problem."

An important element in reducing NOx emissions in aircraft seems to be in developing high strength, high temperature, and low weight materials for engines, Manthey says. "New, advanced materials such as ceramic matrix composites allow engines to burn at higher temperatures," Manthey says. "When engines operate at these high temperatures, they are more efficient.

Other lightweight materials like titanium aluminide and other superalloys allow for more efficient operation of engines because if an engine is lighter, it requires less fuel. These are the technologies we're working with to improve engine performance, which can improve air quality."

Air emission standards can be compared to curbside recycling, says Manthey. It isn't always easy to see visible results that the work you're doing matters, but knowing that you're contributing to a better environment inspires people to continue working. "And that's what we're doing here," she says. "Landing and taking off is when an airplane really affects the air quality in a city. UEET technology will take a good 30 years before you see it widespread in aircraft, so it's important to work on it *now*, before it's too late. This is our future and if we don't fix it in the 21st century, the world in the 22nd century might not be a nice place to live in."

Many people think NASA is only about flying into space, but Manthey's concerns are right here on Earth. "NASA develops technology in aeronautics, not just space. We're taking revolutionary new steps to help aircraft perform better in flight. Air quality affects everyone right here on Earth, and our program will help."

Lesson: Air Pollution

Objective: Students will determine the effects of air pollutants on plants.

Grade Level: 9–12

Subjects: Earth Science, Life Science

Prep Time: 10–30 minutes

Activity Duration: 5 days

Materials Category: Special requirements

National Education Standards:

Science: 1c, 2a, 4f, 7d

Materials:

- Containers (large jars with lids or small fish tanks)
- Small plants
- Air pollutants

Related Links:

Total Ozone Mapping Spectrometer: http://jwocky.gsfc.nasa.gov/

Safeguarding Our Atmosphere: http://www.grc.nasa.gov/WWW/PAO /html/cleaner.htm

Student Sheet(s)

Procedure

Design your own experiment to test the effects of air pollution on plants. Your write-up needs to include the following:

- Purpose
- Hypothesis
- Materials (be specific)
- Procedure
- Data (Make a list of the data you should record.)

Questions

1. What type of seeds or plants was used in your experiment?
2. What pollutants were your plants exposed to?
3. What effect did each pollutant have on each plant?
4. Does the length of time that the seeds were exposed to pollutants affect growth?
5. In what ways might humans be affected by these pollutants?
6. If substances are produced to kill plants (herbicides), what effects could they have on humans who comsume these plants?
7. The average car produces 40 pounds of air pollution for every 1,000 miles driven. If it is driven 15,000 miles each year, how many pounds of air pollution does it produce in 1 year? Explain how you reached your answer.
8. One hundred employees each drive alone 20 miles to work and back each day. Suppose these same employees decide to commute in pairs instead of driving alone. In 2 weeks, they would eliminate 10,000 miles of vehicle travel. They would also save approximately 500 gallons of gasoline and prevent 400 pounds of air pollution. Now, suppose these 100 employees decided to ride four in a car. In 2 weeks:
   a. How many miles of vehicle travel would be eliminated?
   b. How many gallons of gasoline would be saved?
   c. How many pounds of air pollution would be prevented?

Teacher Sheets

Pre-lesson Instructions

- Read to the students, "It's easy to see the air pollution a car makes if you can see smoke coming from its exhaust pipe. But what if you can't see smoke coming from the exhaust pipe? Is the car still making air pollution?"
- Do this prior to class and bring in the white sock or take students outside and demonstrate. Place a clean, white sock over the exhaust pipe of a car.
- Start the engine and let it run for 5 minutes before you remove the sock. Be careful, both the pipe and sock will be hot!
- Ask students, "What did you discover?"
- On most cars, the white sock quickly becomes dirty. Now with the sock off the exhaust pipe, can you see the pollution the car is producing? If it is a fairly new car, you might not see any smoke.

Guidelines

1. Purchase small plants.
2. The most convenient source of pollution is automobile exhaust fumes.
3. For this activity, you may wish to divide students into small lab groups and have them write a procedure. Students can formulate a number of experimental designs from this topic. Have each group explain their proposed procedure. Each group may do the experiment based on their procedure or choose the best two procedures and do them with all the students as a class project.
4. A typical procedure is explained on the second page. (This is not on the Student's Sheet in order not to influence students in designing their own experiment. It can, however, be printed and given to students.)

Extensions

Test the effect of acid rain on plants. (A mixture of 1:3 parts, vinegar and water can be made.) Students can test the pH.

Make a flyer about air pollution. The flyer should include:

- Causes of pollution
- Health effects on living things

- Damage to nonliving things
- How to reduce pollution
- Research sources

Typical Procedure (Several variations possible)

1. Record the type of plant, the heights of the plants, and a written observation of the health of the plants.
2. Place small plants in wide-mouth jars or small terrariums.
3. Water the plants an equal amount. Record amount of water.
4. Put the top on one of the jars and label it "control group."
5. Place another plant in the container close to the exhaust of a running automobile for 1 minute.
6. Close the second container.
7. Repeat steps four and five with the other plants, but vary the exposure time to automobile exhaust. Record the exposure time on the containers.
8. Observe plants after 48 hours and 72 hours.
9. Compare and record results.

Glossary

emission—substances discharged into the air.

flame tubes—the perforated inner tubular 'can' of a gas turbine combustion chamber in which the actual burning occurs.

hypersonic—velocities of Mach number five or above.

matrix composites—more or less continuous matter in which something is embedded.

superalloys—alloy capable of service at high temperatures, usually above 1000 °C.

supersonic—faster than the speed of sound.

LESSON 7

NASA Goes to the Super Bowl

Courtesy of NASA's Aerospace Technology Enterprise

There's nothing more down to Earth than a good old-fashioned game of football, some say. The Super Bowl is the perfect example of simplicity, right? Think again. Advanced technology figures into almost everything we do, and athletics is no exception. At the Super Bowl, you may not see the defense wearing space suits, but there *is* an element of aerospace research being used by just about every player on the field.

Take the helmets, for instance; the padding in the helmets was developed from NASA research, and the plastic in the shell is the same material used by astronauts on space walks. When it comes to the rough-and-tumble game of football, helmets are one of the most crucial pieces of safety equipment. Having well-padded headgear is essential to the safety of athletes involved in the game.

The move to use NASA technology dates back to the late 1960s. NASA Ames Research Center in California developed and tested a foam product that would be commercially marketed as Temper Foam, and was used for the protective padding of airplane seats. Temper Foam is polyurethane plastic foam with several special qualities. It takes the shape of impressed objects, but springs back to its original shape, even if it's been compressed by over 90 percent. It conforms to body shape, evenly distributing weight over the entire contact area for comfort. It also absorbs up to 90 percent of impact shocks and becomes firmer after being subjected to sudden impact. In testing, a 3-inch thick pad could absorb all the energy from a 10-foot fall by an adult.

Besides airplane seats and football helmet padding, Temper Foam has been used for a variety of athletic equipment, including baseball chest protectors and soccer shin guards. In the medical field, the foam is used in wheel chairs to allow people to sit comfortably for longer periods of time.

That was in the 1970s. As time progressed, newer developments in safety foam served both NASA and the sports world well. Temper Foam, while still in existence, has been largely replaced—in airplane seats and in football helmets—with Insullite foam.

While still enormously protective and cushioning, Insullite is lighter weight than Temper Foam, and boasts a more rapid recovery rate. That is, the time it takes for the foam to regain its original size and shape is much shorter. Where Temper Foam might require 10 seconds to regain its shape, Insullite could be ready for its next encounter with a hard object in less than 1 second.

The outer shell of today's football helmets is made from the same plastic material used in astronaut helmets. It's a lightweight polycarbonate with the trade name of Lexan. It helps reduce the impact of a tackle by spreading the force over a greater area of the head. Astronaut helmets feature Lexan in the clear bubble part of the helmet assembly. Lexan is also used in the material for the helmet visors. When you see the astronauts' faces through the helmet visor, you're actually looking through two protective Lexan layers—one clear and one silver-coated.

Football players wear tough, lightweight pads and protective devices. Many of them are made with Kevlar, a material five times stronger than steel. It's the same material used in bulletproof vests... and used to protect space vehicles from stray meteoroid collisions and jet engines from shrapnel.

NASA technology isn't restricted to a football player's uniform. Take a look in the stands. Depending on the weather, you might see a couple more examples of NASA technology used in down-to-Earth ways. Polarized sunglasses with scratch-resistant and ultraviolet ray-resistant lenses were first developed for use in space vehicles. The Mylar thermal blankets that cold sports fans huddle under were first used to insulate in space. Some stadiums (though not the New Orleans facility) have roofs built with the same fabric used in moon suits worn by Apollo astronauts. The Teflon-fiberglass fabric is lightweight, flexible, and waterproof. The material can expand with the heat, contract with the cold, won't catch fire, and allows sunlight to shine through. When it comes to sharing space technology with earthbound groups, NASA and professional football have scored a touchdown!

Lesson: Hey! Watch Out for My Head

Objective: To design a helmet to protect an egg from a fall or collision.

Grade Level: 9–12

Subject(s): Physical Science, Technology

Prep Time: < 10 minutes

Duration: Two class periods

Materials Category: Classroom

National Education Standards:

Science: 3d, 3f, 6a, 6b

Technology: (ITEA) 8, 9, 11

Materials:

- Eggs
- Plastic egg
- Assorted construction materials

Related Link(s):

NASA Site used for derivation of Lesson Plan

An Eggdrop Lander Contest

http://cmex.arc.nasa.gov/Education/MPF_Model/egg.html

Student Sheet(s)

Procedure

1. You will design and build a helmet that will protect a raw egg in a severe fall.

2. Think about the parts of a helmet—hard outside to prevent penetration of foreign objects, and a soft inside to slow the speed of the head (in this case, egg).

3. If the inside of the helmet is too soft, the egg will continue to fall and impact the rigid side of the helmet, which will feel remarkably like the hard ground. If the inside is too hard, the egg will bounce around and possibly crack or break.

4. The outside will determine how the helmet falls. If the outside of the helmet should fail, the contents will be spilled out onto the ground. If the outside causes the helmet to spin and tumble in the air, it will be difficult to insure that the interior is padded the same on all sides.

5. The egg must be able to get in and out of the helmet. This process may have only two steps and not take more than 30 seconds to complete.

6. The egg must be able to see with the helmet on. Do not cover its face.

7. Listen carefully to your teacher for information regarding specific design rules and how your helmet will be judged.

8. Design and build the exterior. Make sure it is big enough to hold an egg and its protective material.

9. Design the internal structure of the helmet, called the "Egg Protection System" (EPS).

10. The egg must look stylish in its helmet. The helmet must be colorful and neatly constructed.

11. Use the plastic egg to test your helmet.

12. On test day, draw a face on your egg.

13. Strap your egg in the EPS, and let the test begin.

Teacher Sheets

Pre-lesson Instructions

- Encourage creative ideas. One student did a project in which they placed the airbags (balloons) on the inside of the "eggtainer" to protect the egg. It was virtually "bulletproof" and was even slammed into the ground with no damage to the egg.
- Even if you have never had a problem with eggs and students, always be aware of temptations!

Guidelines

1. Read the NASAexplores article, "NASA Goes to the Super Bowl."
2. Put constraints on design that challenge the level of your students and the materials available (weight of helmet, overall size of helmet, how the results will be judged, distance from target), and explain these design criteria to the class.
3. Have students assemble test helmets for experimenting. They should be allowed some testing time using plastic eggs. NO parachutes! Imagine playing football with a parachute attached to your helmet.
4. Have the students design the internal structure, called the "Egg Protection System" (EPS), of the helmet.
5. Attempt to get as much altitude as possible for your drops.
6. You provide the eggs, and have the students load them in the Test Area. You don't want eggs dropped in the classroom.
7. It is best to have someone impartial make the actual drop, but encourage teams to be involved in the preparations.
8. As the helmets land, have a group of judges measure the distance from the center of the target area, and make the determination as to the condition of the egg.

Discussion/Wrap-Up

- Use increasing drop heights to determine the winner. The egg helmet that survives the highest drop will be the winner.
- What designs worked the best? Why?

Glossary

polycarbonate—any of a family of thermoplastics characterized by high-impact strength, light weight, and flexibility, and used as a shatter-resistant substitute for glass.

polyurethane—any of various resins, widely varying in flexibility, used in tough chemical-resistant coatings, adhesives, and foams.

LESSON 8

Keeping Comfortable with a NASA Wardrobe

Courtesy of NASA's Human Exploration and Development of Space Enterprise. Excerpted.

Moon wear provided more than just athletic shoe technology. Because of temperature extremes on the lunar surface, astronauts needed to maintain their body temperatures in situations ranging from very hot to ultracold.

Aluminized mylar (mylar is the shiny material in foil balloons), originally developed for NASA Goddard Space Flight Research Center to make satellites more reflective, and for space suit insulation, has been adapted to create jackets and ski parkas. The reflective capabilities are used to retain body heat and provide a barrier from cold and hot temperatures.

Space suits featured heat-absorbing gel packs that slip into insulated pockets of the astronauts' garments, positioned near parts of the body where heat transfer is most efficient. Gel packets, which last about an hour and are easily replaced, are invaluable during space walks. Runners, joggers, and any other athlete on Earth whose performance may be affected by hot weather can wear cooling headbands, wristbands and running shorts with gel pack pockets. Gel packs can have nonathletic uses as well: hot and cold compresses for sore muscles and temperature control for sports spectators are just two possibilities.

Lesson: Keeping Cool

Objective: To learn about the functionality of the cooling process in relation to space suit designs.

Grade Level: 9–12

Subject: Science

Prep Time: 10–30 minutes

Activity Duration: 50-minute class period

Materials Category: Special requirements

National Education Standards

Science: 3f, 4f, 6a

Technology ITEA: 14a, 16c, 19f

Materials:

- Two coffee cans with plastic lids
- Four meters of aquarium tubing
- Two buckets
- Two thermometers
- Duct tape
- Water (solid and liquid)
- Heat source (light bulb and fixture)
- Single-hole punch
- Flood light and fixture

Related Link:

Suited for Spacewalking: http://spacelink.nasa.gov/Instructional.Materials/NASA.Educational.Products/Suited.For.Spacewalking/Suited.for.Spacewalking.pdf

Student Sheet(s)

Background

It is not sufficient for an astronaut just to be protected from the hazards of the environment in which he or she is trying to work. One of the most important hazardous conditions is temperature. Suit insulation technologies protect the astronaut from extreme high and low temperatures of the space environment. However, the same insulation technology also works to keep heat released by the astronaut's body inside the suit. To get an idea of what this is like, imagine walking around in summer wearing a plastic bag.

In Space Shuttle Extravehicular Mobility Units (EMUs), the cooling system consists of a network of small diameter water circulation tubes that are held close to the body by a Spandex® body suit. Heat released by the astronaut's body is transferred to the water where it is carried to a refrigeration unit in the suit's backpack. The water runs across a porous metal plate that is exposed to the vacuum of outer space on the other side. Small amounts of water pass through the pores where it freezes on the outside of the plate. As additional heated water runs across the plate, the heat is absorbed by the aluminum and is conducted to the exposed side. There the ice begins to sublimate, or turn directly into water vapor, and disperses in space. Sublimation is a cooling process. Additional water passes through the pores and freezes as before. Consequently, the water flowing across the plate has been cooled again and is used to recirculate through the suit to absorb more heat.

Supplementing the EMU cooling system is an air-circulation system that draws perspiration from the suit into a water separator. The water is added to the cooling water reservoir while the drier air is returned to the suit. The cooling system and the air circulation system work together to provide a comfortable internal working environment. The wearer of the suit controls the operating rates of the system through controls on the Display and Control Module mounted on the EMU chest. The system can eliminate 40–60 percent of stored body heat.

Stevie Roper of Waynesville, North Carolina was born without sweat glands, a condition known as hypohidrotic ectodermal dysplasia (HED). People with HED are in constant threat of heat exhaustion or stroke because they have no way to cool themselves. NASA space suit technology has a way to help: a special cooling undergarment was developed in 1968 at NASA's Ames Research Center for astronauts to wear with a space suit. The system circulates a fluid through a series of tubes next to the body; the fluid picks up body heat and carries it away. Stevie's aunt, Sarah Ann Moody of Hampton, Virginia, was referred by NASA's Langley Research Center to Life Support Systems, Inc. (LSSI), a manufacturer of personal cooling systems. With her perseverance, enough money was raised to buy a special version of the cooling suit. She has since founded the HED Foundation to help provide the suits to children in the United States and other countries.

The suits make it possible for the children to be active, improving their lives dramatically. Stevie Roper is able to have a more normal life by wearing the space-derived cooling suit.

Procedure

1. Punch a hole near the bottom of the wall of a metal coffee can. The hole should be large enough to pass aquarium tubing through.

2. Punch a second hole in the plastic lid of the can so that tubing can pass through it as well.

3. Punch another hole in the center of the lid so that a thermometer will fit snugly into it.

4. Finally, punch a hole in the center of the second coffee can lid for another thermometer.

5. Loosely coil the aquarium tubing, and place it inside the first coffee can. Use bits of tape to hold the coils to the walls and to keep them spread out evenly. Pass the lower end of the tube through the hole in the can wall and the upper end through the outer hole in the lid. The lower tube should extend to the catch bucket that will be placed below the can. The upper end will have to reach to the bottom of the ice water bucket. That bucket will be elevated above the can.

6. Insert a thermometer into each can.

7. Place the two cans on a tabletop. Direct the light from a strong light bulb or floodlight to fall equally on the two cans. The light should be no more than about 25 centimeters away from the cans.

8. Fill a bucket with ice water, and elevate it above the two cans. You can color the water with food coloring to increase its visibility in the siphons.

9. Place the catch bucket below the two cans.

10. Turn on the light. Observe and record the temperatures on the two thermometers. After 2 minutes, again observe and record the temperatures.

11. Place the upper end of the aquarium tubing into the ice water and suck on the other end of the tube to start a siphon flowing. Let the water pour into the catch bucket.

12. Observe and record the temperature of the two cans at 30-second intervals for 10 minutes.

13. Plot the temperature data on graph paper, using a solid line for the can that held the ice water and a dashed line for the other can. Construct the graph so that the temperature data are along the Y-axis and those for time along the X-axis. Compare the slope of the plots for the two cans.

Questions

1. How can the flow of icy water be controlled?

2. Suggest a way to maintain a constant temperature inside the can with the tubing.

3. What would happen if you moved the light source closer to the can?

4. Explain how a liquid cooling garment could be constructed that could operate continuously without siphons and buckets of ice water that eventually run out.

5. Can you think of any practical application of the EMU suit on Earth?

Extra Credit

Design a liquid cooling garment out of long underwear or Spandex® running tights.

Teacher Sheets

Background

Astronauts out on extravehicular activity are in a constant state of exertion. Body heat released from this exertion can quickly build up inside a space suit, leading to heat exhaustion. Body heat is controlled by a liquid cooling garment made from stretchable spandex fabric and laced with small diameter plastic tubes that carry chilled water. The water is circulated around the body. Excess body heat is absorbed into the water and carried to the suit's backpack, where it runs along a porous metal plate that permits some of the heat to escape into outer space. The water instantly freezes on the outside of the plate and seals the pores. More water circulates along the back of the plate. Heat in the water is conducted through the metal to melt the ice into water vapor. In the process, the circulating water is chilled. The process of freezing and thawing continues constantly at a rate determined by the heat output of the astronaut.

This activity demonstrates how chilled water can keep a metal can from heating up even when exposed to the strong light of a floodlight.

Pre-lesson Instructions

- Collect coffee cans.
- Holes in the cans may be made for the students using a metal punch or drill.

Guidelines

- Read the 9–12 NASAexplores article, "NASA Technology in Your Wardrobe." Discuss how NASA technology has made an impact on clothing for the general public.

Extensions

- Ask each student to place one bare arm inside a tall kitchen plastic garbage bag and wrap the bag snugly around the arm. The bag represents the restraint layer of a space suit. After a minute or two, ask the students to compare how their covered arm feels compared to the uncovered arm? Why is there a difference? How would you feel if your entire body was covered like this?
- Make small bags of various materials to test their insulating properties. Slip a thermometer into each bag and measure the bags' temperature rise when exposed to a heat source such as a floodlight or sunlight. Try using fabrics,

paper, aluminum foil, and plastics as well as commercial insulating materials such as rock wool and cellulose. Also, experiment with multilayered materials. Compare the bulk and weight of different insulators with their effectiveness. What criteria must space suit designers use in evaluating space suit insulation?

Glossary

hydrodynamic—the branch of dynamics that studies the motion produced in fluids by applied forces.

temper—a vague term describing the relative condition of the hardness and mechanical properties of a substance.

LESSON 9

Water: It's Not Just for Drinking

Question: In space, what do water and air have in common? Answer: They don't exist. If you want them, you have to take them with you when you launch from Earth.

Water and air are essential for life. Making sure that supplies are pure and plentiful is a high priority. On the International Space Station (ISS), water is a precious commodity, so it's never wasted. A crew of four astronauts uses about 18,000 kilograms (40,000) pounds of water in one year.

NASA researchers have worked hard to develop new ways to do traditional activities so that they will require less water, or in some cases, none at all. For example, astronauts use a special type of toothpaste so that rinsing isn't necessary. They also use shampoo that doesn't need rinsing. The toilets flush with air instead of water. When bathing, astronauts clean their bodies with a moistened washcloth.

Food can also be prepared without the use of water. Some meals are rehydrated, but many more are served much like TV dinners—by reheating precooked items.

While water doesn't exist in space, it *is* created as a human byproduct, and must be removed. Perspiration and breathing create moisture in the air; if it's not removed, it will collect on and damage equipment. It can also make breathing uncomfortable.

Conservation isn't enough, though. Existing water must be recycled or reprocessed so that it can be used again. Purified water, urine, and perspiration don't sound very appetizing, do they? Astronauts compare this to how water is recycled in nature. Waste products filter through the ground water system, and the water flows into rivers, where it is collected for use in our communities. Nature does an amazing job of purifying water—and NASA has done a great job of imitating nature. The Environmental Control and Life Support System (ECLSS) is currently in development by NASA at Marshall Space Flight Center in Alabama. The ECLSS consists of seven systems that are responsible for: water recycling and purification, oxygen generation and purification, air conditioning, and atmospheric pressure. These systems interact to provide a habitable environment for the flight crew in the crew compartment in addition to cooling or heating various orbiter systems or components. By the time water has passed through the rigorous purification standards of ECLSS, it will be cleaner than the tap water we use on Earth.

Once water purification becomes a well-developed procedure in space, it can be used on Earth where clean water is a concern. In cold areas like Antarctica where normal biological processes don't work, a system like ECLSS could provide people living there with pure water that would otherwise be unavailable.

Is Your Water Clean?

Objective: To test water quality for possible contamination.

Subject(s): Chemistry, Biology, Earth Science

Prep Time: > 30 minutes

National Education Standards:

Science: 7a, 7c, 7d, 7e

Grade Level: 9–12

Duration: One Class Period

Materials Category: Special

Materials:

- Goggles (per student)
- Rubber gloves
- Phosphate test kit and directions
- Coliform test kit and directions
- pH paper
- Data Sheets (See the Student Sheets.)
- Alcohol (for hand cleaning)
- Electronic Sensor (electrodes to test for current conduction through water)
- Microscopes
- Glass containers to hold water samples (five per group)

Related Link(s):

Ground Water Quality—Contamination Menu
http://www.epa.gov/seahome/groundwater/src/quality2.htm

Supporting NASAexplores Article(s):

Water: It's Not Just for Drinking
http://www.nasaexplores.com/lessons/02-054/9-12_article.html

Student Sheet(s)

Materials

- Goggles (per student)
- Rubber gloves
- Phosphate test kit and directions
- Coliform test kit and directions
- pH paper
- Data Sheets
- Alcohol (for hand cleaning)
- Electronic Sensor (electrodes to test for current conduction through water)
- Microscopes
- Glass containers to hold water samples (five per group)

Procedure

1. Choose one of the water samples to test. Be sure each member of your group gets a different sample to test.

2. Test the odor of your water sample by smelling it. Record your data on the Data Sheet. Write any conclusions you can make from the scent of the water in the conclusion section of the Data Sheet. Pure water has no odor.

3. Test the clarity/color of your water. Record your data and any conclusions you can make. Turbidity refers to the opaqueness or cloudiness of water—the muddier the water, the more turbid it is. Suspended solids in the water create turbidity, which can be measured by how much light is blocked or absorbed.

4. Carefully read the directions for the phosphate test kit. Perform the phosphate test on your water sample. Record your data and any conclusions you

can make. Phosphorus is one of the key elements necessary for growth of plants and animals. Rainfall can cause varying amounts of phosphates to wash from farm soils into nearby waterways. Phosphate will stimulate the growth of plankton and aquatic plants that provide food for fish. This increased growth may cause an increase in the fish population and improve the overall water quality. However, if an excess of phosphate enters the waterway, algae and aquatic plants will grow wildly, choke up the waterway and use up large amounts of oxygen. This condition is known as eutrophication.

5. Using the pH paper, test the acidity of your water sample. Record your data and any conclusions you can make. A pH range of 6.0 to 9.0 appears to provide protection for the life of freshwater fish and bottom-dwelling invertebrates.

6. Carefully read the directions for the coliform test kit. Perform the coliform test on your water sample. Record your data and any conclusions you can make.

7. Look at a drop of your water sample under a microscope. Look for bacteria and other "small creatures" in your water sample. Record your data and any conclusions you can make.

8. With the assistance of the teacher, check for total dissolved solids, like salt, using a conductivity tester. **Be careful to avoid electric shock!** Record your data and any conclusions you can make.

Conclusion

Write a conclusion paragraph explaining why drinking water should be tested.

Teacher Sheets

Pre-lesson Instructions

- Collect three large containers of water from three different sources such as: water from a local river, tap water, and water from an irrigation ditch. Water from a local stream or river needs to be collected in clean, clear containers. (Remember that in order not to kill off microorganisms, containers cannot be kept tightly sealed for any length of time!)
- Go over the instructions for the kits and pH paper with the students.
- Stress to students that they must wear safety goggles and gloves at all times when working with chemicals and unknown water sources.

- Remind students to wash hands after contact with any unknown water sources.

Background Information

Drinking water comes from a variety of sources. Some of the water comes from water purification plants. Some comes from underground sources. Due to the diversity of the sources, the water we drink can differ greatly in quality and healthiness. The study of water is limnology. This involves physical, chemical, and biological conditions.

Physical conditions refer to water temperature, stream velocity, and turbidity (clarity). Chemical conditions refer to the chemical make up of the water. This includes the amounts of dissolved oxygen, phosphate and nitrate. Biological conditions refer to organisms supported in the water such as bacteria, plankton, and fish.

Guidelines

1. Distribute and read the 9–12 NASAexplores article, "Water: It's Not Just for Drinking." Discuss what would be important to do if the astronauts needed to recycle water (removal of pollutants). Emphasize the need for the proper pH level (U.S. standards call for a pH of 6.5 to 8.5).

2. Distribute the Student Sheets and the materials.

3. Go over safety instructions (that is, any water spilled must be cleaned up immediately to prevent slips and falls). None of the materials are dangerous if proper lab safety guidelines are followed.

4. Students should work in groups of three. Each student should have his/her own sample of water to test. Be sure each group member has a different sample to test.

5. Ask students, "What are some indicators of water quality?" Write their answers on the board. Explain to students that they will be performing several tests to determine the quality of a sample of water.

6. Students will test each sample for:

   a. odor
   b. clarity/color
   c. phosphates
   d. pH
   e. fecal coliforms

f. observe through a microscope for bacterial forms

g. total dissolved solids, like salt, using a conductivity tester (with assistance from the teacher). Be careful to avoid electric shock!

Discussion/Wrap-Up

Go over the results of the tests with students orally and explain what each test might indicate.

Bad odor—could indicate sewage pollution, algae. Chlorine odor could indicate treatment from a sewage treatment plant.

Clarity/color—poor clarity could indicate dissolved (suspended) solids, like silt or soil in the water.

Phosphates—if phosphates are present, they could indicate the presence of fertilizers, wastewater (detergents, sewage, etc.), or industrial discharges. These lead to algae blooms and plant blooms that consume carbon dioxide (CO_2) and kill everything in the water.

Acidity (pH)—most biological systems use pH at approximately 7.1. A low pH (acidic, below 5) or high (alkaline, above 9) may kill eggs, larvae, nymphs, hatchlings, etc., as well as leach toxic heavy metals from soils and rock.

Fecal coliforms—these are bacteria derived from human feces, mainly *E. coli*. See the direction kit for levels. (High levels indicate contamination, possibly sewage being too close to the water supply.)

Microscope observations—some bacteria are normal and harmless. But, it is interesting to see what kind of "critters" are in the water we drink.

Total dissolve solids (conductivity)—if your sample is conductive—or the degree to which it is conductive—it tells you the degree of particles present in your sample. For example, the more salts the more conductive the sample.

Extensions

Have students bring in tap water from their homes to test.

Test samples of water that are contaminated, and then run them through a water purification system. Retest the water to see the effectiveness of the purification system.

National Education Standards
(Standards are for grades 9–12 unless otherwise noted)

Science

National Science Education Standards

NSTA National Science Teachers Association

1. Unifying Concepts: *K–12 Targets*
   a. Systems, order, and organization
   b. Evidence, models, and explanation
   c. Change, constancy, and measurement
   d. Evolution and equilibrium
   e. Form and function
2. Science as Inquiry: *9–12 Targets*
   a. Abilities necessary to do scientific inquiry
      1. Identify questions and concepts that guide scientific investigations
      2. Design and conduct scientific investigations
      3. Use technology and mathematics to improve investigations and communications
      4. Formulate and revise scientific explanations and models using logic and evidence
      5. Recognize and analyze alternative explanations and models
      6. Communicate and defend a scientific argument
   b. Understandings about scientific inquiry
3. Physical Science: *9–12 Targets*
   a. Structure of atoms
   b. Structure and properties of matter
   c. Chemical reactions

d. Motion and forces

e. Conservation of energy and increase in disorder

f. Interaction of energy and matter

4. Life Science: *9–12 Targets*

a. The cell

b. Molecular basis of heredity

c. Biological evolution

d. Interdependence of organisms

e. Matter, energy, and organization of living systems

f. Behavior of organisms

5. Earth and Space Science: *9–12 Targets*

a. Energy in the earth systems

b. Geochemical cycles

c. Origin & evolution of the earth system

d. Origin & evolution of the universe

6. Science and Technology: *9–12 Targets*

a. Abilities of technological design

b. Understanding about science and technology

7. Science in Personal and Social Perspectives: *9–12 Targets*

a. Personal and community health

b. Population growth

c. Natural resources

d. Environmental quality

e. Natural and human-induced hazards

f. Science and technology in local, national, and global challenges

8. History and Nature of Science: *9–12 Targets*

    a. Science as a human endeavor

    b. Nature of scientific knowledge

    c. Historical perspectives

Mathematics

Principles and Standards for School Mathematics

NCTM National Council of Teachers of Mathematics

Numbers and Operations

1. Understand numbers, ways of representing numbers, relationships among numbers, and number systems: *9–12 Targets*

    a. Develop a deeper understanding of very large and very small numbers and of various representations of them

    b. Compare and contrast the properties of numbers and number systems, including the rational and real numbers, and understand complex numbers as solutions to quadratic equations that do not have real solutions

    c. Understand vectors and matrices as systems that have some of the properties of the real-number system

    d. Use number-theory arguments to justify relationships involving whole numbers

2. Understand meanings of operations and how they relate to one another: *9–12 Targets*

    a. Judge the effects of such operations as multiplication, division, and computing powers and roots on the magnitudes of quantities

    b. Develop an understanding of properties of, and representations for, the addition and multiplication of vectors and matrices

    c. Develop an understanding of permutations and combinations as counting techniques

3. Compute fluently and make reasonable estimates: *9–12 Targets*

a. Develop fluency in operations with real numbers, vectors, and matrices, using mental computation or paper-and-pencil calculations for simple cases and technology for more complicated cases

b. Judge the reasonableness of numerical computations and their results

Algebra

4. Understand patterns, relations, and functions: *9–12 Targets*

a. Generalize patterns using explicitly defined and recursively defined functions

b. Understand relations and functions and select, convert flexibly among, and use various representations for them

c. Analyze functions of one variable by investigating rates of change, intercepts, zeros, asymptotes, and local and global community

d. Understand and perform transformations such as arithmetically comb ining, composing, and inverting commonly used functions, using technology to perform such operations on more-complicated symbolic expressions

e. Understand and compare the properties of classes of functions, including exponential, polynomial, rational, logarithmic, and periodic functions

f. Interpret representations of functions of two variables

5. Represent and analyze mathematical situations and structures using algebraic symbols: *9–12 Targets*

a. Understand the meaning of equivalent forms of expressions, equations, inequalities, and relations

b. Write equivalent forms of equations, inequalities, and systems of equations, and solve them with fluency—mentally or with paper and pencil in simple cases and using technology in all cases

c. Use symbolic algebra to represent and explain mathematical relationships

d. Use a variety of symbolic representations, including recursive and parametric equations, for functions and relations

e. Judge the meaning, utility, and reasonableness of the results of symbol manipulations, including those carried out by technology

6. Use mathematical models to represent and understand quantitative relationships

   a. Identify essential quantitative relationships in a situation and determine the class or classes of functions that might model the relationships

   b. Use symbolic expressions, including iterative and recursive forms, to represent relationships arising from various contexts

   c. Draw reasonable conclusions about a situation being modeled

7. Analyze change in various contexts

   a. Approximate and interpret rates of change from graphical and numerical data

Geometry

8. Analyze characteristics and properties of two- and three-dimensional geometric shapes and develop mathematical arguments about geometric relationships: *9–12 Targets*

   a. Analyze properties and determine attributes of 2- and 3-dimensional objects

   b. Explore relationships (including congruence and similarity) among classes of 2- and 3-dimensional geometric objects, make and test conjectures about them, and solve problems involving them

   c. Establish the validity of geometric conjectures using deduction, prove theorems, and critique arguments made by others

   d. Use trigonometric relationships to determine lengths and angle measures

9. Specify locations and describe spatial relationships using coordinate geometry and other representational systems: *9–12 Targets*

   a. Use Cartesian coordinates and other coordinate systems, such as navigational, polar, or spherical systems, to analyze geometric situations

   b. Investigate conjectures and solve problems involving 2- and 3-dimensional objects represented with Cartesian coordinates

10. Apply transformations and use symmetry to analyze mathematical situations: *9–12 Targets*

a. Understand and represent translations, reflections, rotations, and dilations of objects in the plane by using sketches, coordinates, vectors, function notation, and matrices

b. Use various representations to help understand the effects of simple transformations and their compositions

11. Use visualization, spatial reasoning, and geometric modeling to solve problems: *9–12 Targets*

a. Draw and construct representations of 2- and 3-D geometric objects using a variety of tools

b. Visualize 3-D objects from different perspectives and analyze their cross sections

c. Use vertex-edge graphs to model and solve problems

d. Use geometric models to gain insight into, and answer questions in, other areas of mathematics

e. Use geometric ideas to solve problems in, and gain insights into, other disciplines and other areas of interest such as art and architecture

Measurement

12. Understand measurable attributes of objects and the units, systems, and processes of measurement: *9–12 Targets*

a. Make decisions about units and scales that are appropriate for problem situations involving measurement

13. Apply appropriate techniques, tools, and formulas to determine measurements: *9–12 Targets*

a. Analyze precision, accuracy, and approximate error in measurement situations

b. Understand and use formulas for the area, surface area, and volume of geometric figures, including cones, spheres, and cylinders

c. Apply informal concepts of successive approximation, upper and lower bounds, and limit in measurement situations

d. Use unit analysis to check measurement computations

Data Analysis and Probability

14. Formulate questions that can be addressed with data and collect, organize, and display relevant data to answer them: *9–12 Targets*

    a. Understand the differences among various kinds of studies and which types of inferences can be legitimately drawn from each

    b. Know the characteristics of well-designed studies, including the role of randomization in surveys and experiments

    c. Understand the meaning of measurement data and categorical data, of univariate and bivariate data, and of the term variable

    d. Understand histograms, parallel box plots, and scatterplots and use them to display data

    e. Compute basic statistics and understand the distinction between a statistic and a parameter

15. Select and use appropriate statistical methods to analyze data: *9–12 Targets*

    a. For univariate measurement data, be able to display the distribution, describe its shape, and select and calculate summary statistics

    b. For bivariate measurement data, be able to display a scatterplot, describe its shape, and determine regression coefficients, regression equations, and correlation coefficients using technological tools

    c. Display and discuss bivariate data where at least one variable is categorical

    d. Recognize how linear transformations of univariate data affect shape, center, and spread

    e. Identify trends in bivariate data and find functions that model the data or transform the data so they can be modeled

16. Develop and evaluate inferences and predictions that are based on data: *9–12 Targets*

    a. Use simulations to explore the variability of sample statistics from a known population and to construct sampling distributions

    b. Understand how sample statistics reflect the values of population parameters and use sampling distributions as the basis for informal inference

c. Evaluate published reports that are based on data by examining the design of the study, the appropriateness of the data analysis, and the validity of conclusions

d. Understand how basic statistical techniques are used to monitor process characteristics in the workplace

17. Understand and apply basic concepts of probability: *9–12 Targets*

a. Understand the concepts of sample space and probability distribution and construct sample spaces and distributions in simple cases

b. Use simulations to construct empirical probability distributions

c. Compute and interpret the expected value of random variables in simple cases

d. Understand the concepts of conditional probability and independent events

e. Understand how to compute the probability of a compound event

18. Problem Solving: *K–12 Targets*

a. Build new mathematical knowledge through problem solving

b. Solve problems that arise in mathematics and in other contexts

c. Apply and adapt a variety of appropriate strategies to solve problems

d. Monitor and reflect on the process of mathematical problem solving

19. Reasoning and Proof: *K–12 Targets*

a. Recognize reasoning and proof as fundamental aspects of mathematics

b. Make and investigate mathematical conjectures

c. Develop and evaluate mathematical arguments and proofs

d. Select and use various types of reasoning and methods of proof

20. Communication: *K–12 Targets*

a. Organize and consolidate their mathematical thinking through communication

b. Communicate their mathematical thinking coherently and clearly to peers, teachers, and others

c. Analyze and evaluate the mathematical thinking and strategies of others

d. Use the language of mathematics to express mathematical ideas precisely

21. Connections: *K–12 Targets*

a. Recognize and use connections among mathematical ideas

b. Understand how mathematical ideas interconnect and build on one another to produce a coherent whole

c. Recognize and apply mathematics in contexts outside of mathematics

22. Representation: *K–12 Targets*

a. Create and use representation to organize, record, and communicate mathematical ideas

b. Select, apply, and translate among mathematical representations to solve problems

c. Use representations to model and interpret physical, social, and mathematical phenomena

Technology—ISTE

Technology Foundation Standards for All Students

ISTE International Society for Technology in Education

Basic Operations and Concepts

1. Students demonstrate a sound understanding of the nature and operation of technology systems.

2. Students are proficient in the use of technology: *9–12 Targets*

   a. Make informed choices among technology systems, resources, and services

Social, Ethical, and Human Issues

3. Students understand the ethical, cultural, and societal issues related to technology.

4. Students practice responsible use of technology systems, information, and software.

5. Students develop positive attitudes toward technology uses that support lifelong learning, collaboration, personal pursuits, and productivity: *9–12 Targets*

   a. Identify capabilities and limitations of contemporary and emerging technology resources and assess the potential of these systems and services address personal, lifelong learning, workplace needs

   b. Make informed choices among technology systems, resources, and services

   c. Analyze advantages and disadvantages of widespread use and reliance on technology in the workplace and in society as a whole

   d. Demonstrate and advocate for legal and ethical behaviors among peers, family, and community regarding the use of technology and information

Technology Productivity Tools

6. Students use technology tools to enhance learning, increase productivity, and promote creativity

7. Students use productivity tools to collaborate in constructing technology-enhanced models, prepare publications, and produce other creative works: *9–12 Targets*

   a. Use technology tools and resources for managing and communicating personal/professional information (e.g., finances, schedules, addresses, purchases, correspondence)

   b. Investigate and apply expert systems, intelligent agents, and simulations in real-world situations

Technology Communications Tools

8. Students use telecommunications to collaborate, publish, and interact with peers, experts, and other audiences

9. Students use a variety of media and formats to communicate information and ideas effectively to multiple audiences: *9–12 Targets*

   a. Use technology tools and resources for managing and communicating personal/professional information (e.g., finances, schedules, addresses, purchases, correspondence)

b. Routinely and efficiently use online information resources to meet needs for collaboration, research, publications, communications, and productivity

c. Select and apply technology tools for research, information analysis, problem-solving, and decision-making in content learning

d. Collaborate with peers, experts, and others to contribute to a content-related knowledge base by using technology to compile, synthesize, produce, and disseminate information, models, and other creative works

Technology Research Tools

10. Students use technology to locate, evaluate, and collect information from a variety of sources

11. Students use technology tools to process data and report results

12. Students evaluate and select new information resources and technological innovations based on the appropriateness for specific tasks: *9–12 Targets*

    a. Evaluate technology-based options, including distance and distributed education, for lifelong learning

    b. Routinely and efficiently use online information resources to meet needs for collaboration, research, publications, communications, and productivity

    c. Select and apply technology tools for research, information analysis, problem-solving, and decision-making in content learning

    d. Investigate and apply expert systems, intelligent agents, and simulations in real-world situations

    e. Collaborate with peers, experts, and others to contribute to a content-related knowledge base by using technology to compile, synthesize, produce, and disseminate information, models, and other creative works

Technology Problem-Solving and Decision-Making Tools

13. Students use technology resources for solving problems and making informed decisions.

14. Students employ technology in the development of strategies for solving problems in the real world: *9–12 Targets*

a. Routinely and efficiently use online information resources to meet needs for collaboration, research, publications, communications, and productivity

b. Investigate and apply expert systems, intelligent agents, and simulations in real-world situations

c. Collaborate with peers, experts, and others to contribute to a content-related knowledge base by using technology to compile, synthesize, produce, and disseminate information, models, and other creative works

Technology—ITEA

Standards for Technological Literacy: Content for the Study of Technology

ITEA International Technology Education Association

1. Characteristics and scope of technology: *9–12 Targets*

   a. Nature of technology

   b. Rate of technological diffusion

   c. Goal-directed research

   d. Communication of technology

2. Core concepts of technology: *9–12 Targets*

   a. Systems

   b. Resources

   c. Requirements

   d. Optimization and trade-offs

   e. Processes

   f. Controls

3. Relationships among technologies and other fields: *9–12 Targets*

   a. Technology transfer

   b. Innovation and invention

c. Knowledge protection and patents

d. Technological knowledge and advances of science and mathematics

4. Cultural, social, economic, and political effects: *9–12 Targets*

a. Rapid or gradual changes

b. Reduce resource use

c. Monitor environment

d. Alignment of natural and technological processes

e. Reduce negative consequences of technology

f. Decision and trade-offs

5. Effects of technology on environment: *9–12 Targets*

a. Conservation

b. Reduce resource use

c. Monitor environment

d. Alignment of natural and technological processes

e. Reduce negative consequences of technology

f. Decision and trade-offs

6. Role of society in the development and use of technology: *9–12 Targets*

a. Different cultures and technologies

b. Development decisions

c. Factors affecting designs and demands of technology

7. Influence of technology on history: *9–12 Targets*

a. Evolutionary development

b. Dramatic changes in society

c. History of technology

d. Early technological history

e. The Iron Age

f. The Middle Age

g. The Renaissance

h. The Industrial Revolution

i. The Information Age

8. Attributes of design: *9–12 Targets*

a. The design process

b. Design problems are usually not clear

c. Designs need to be refined

d. Requirements

9. Engineering design: *9–12 Targets*

a. Design principles

b. Influence of personal characteristics

c. Prototypes

d. Factors in engineering design

10. Role of troubleshooting, research and development, inventions and innovation, and experimentation in problem solving: *9–12 Targets*

a. Research and development

b. Researching technological problems

c. Not all technology problems can be solved

d. Multidisciplinary approach

11. Apply the design process: *9–12 Targets*

a. Identify a design problem

b. Identify criteria and constraints

c. Refine the design

d. Evaluate the design

e. Develop a product or system using quality control

f. Reevaluate final solutions

12. Use and maintain technological products and systems: *9–12 Targets*

    a. Document and communicate processes and procedures

    b. Diagnose a malfunctioning system

    c. Troubleshoot and maintain systems

    d. Operate and maintain systems

    e. Use computers to communicate

13. Assess impact of products and systems: *9–12 Targets*

    a. Collect information and judge quality

    b. Synthesize data to draw conclusions

    c. Employ assessment techniques

    d. Design forecasting techniques

14. Medical technologies: *9–12 Targets*

    a. Medical technologies for prevention and rehabilitation

    b. Telemedicine

    c. Genetic therapeutics

    d. Biochemistry

15. Agricultural and related biotechnologies: *9–12 Targets*

    a. Agricultural products and systems

    b. Biotechnology

c. Conservation

d. Engineering design and management of ecosystems

16. Energy and power technologies: *9–12 Targets*

a. Law of conservation of energy

b. Energy sources

c. Second Law of Thermodynamics

d. Renewable and nonrenewable forms of energy

e. Power systems are a source, a process, and a load

17. Information and communication technologies: *9–12 Targets*

a. Parts of information and communication systems

b. Information and communication systems

c. The purpose of information and communication technology

d. Communication systems and sub-systems

e. Many ways of communicating

f. Communicating through symbols

18. Transportation technologies: *9–12 Targets*

a. Relationship of transportation and other technologies

b. Intermodalism

c. Transportation of services and methods

d. Positive and negative impacts of transportation systems

e. Transportation processes and efficiency

19. Manufacturing technologies: *9–12 Targets*

a. Servicing obsolescence

b. Durable or non-durable goods

c. Manufacturing systems

d. Interchangeability of parts

e. Chemical technologies

f. Marketing of products

20. Construction technologies: *9–12 Targets*

a. Infrastructure

b. Construction processes and procedures

c. Requirements

d. Prefabricated materials

Geography

Geography for Life: National Geography Standards

NGS National Geographic Society

The World in Spatial Terms

1. How to use maps and other geographic representations, tools, and technologies to acquire, process, and report information from a spatial perspective: *9–12 Targets*

   a. How to use maps and other graphic representations to depict geographic problems

   b. How to use technologies to represent and interpret Earth's physical and human systems

   c. How to use geographic representatives and tools to analyze, explain, and solve geographic problems

2. How to use mental maps to organize information about people. Places, and environments in a spatial context: *9–12 Targets*

   a. How to use mental maps of physical and human features of the world to answer complex geographic questions

   b. How mental maps reflect the human perception of places

   c. How mental maps influence spatial and environmental decision-making

3. How to analyze the spatial organization of people, places, and environments on earth's surface: *9–12 Targets*

    a. The generalizations that describe and explain spatial interaction

    b. The models that describe patterns of spatial organization

    c. The spatial behavior of people

    d. How to apply concepts and models of spatial organization to make decisions

Places and Regions

4. The Physical and Human Characteristics of Places: *9–12 Targets*

    a. The meaning and significance of place

    b. The changing physical and human characteristics of places

    c. How relationships between humans and the physical environment lead to the formation of places and to a sense of personal and community identity

5. That people create regions to interpret earth's complexity: *9–12 Targets*

    a. How multiple criteria can be used to define a region

    b. The structure of regional systems

    c. The ways in which physical and human regional systems are interconnected

    d. How to use regions to analyze geographic issues

6. How culture and experience influence people's perceptions of places and regions: *9–12 Targets*

    a. Why places and regions serve as symbols for individuals and society

    b. Why different groups of people within a society view places and regions differently

    c. How changing perceptions of places and regions reflect cultural change

7. The physical processes that shape the patterns of earth's surface: *9–12 Targets*

a. The dynamics of the four basic components of Earth's physical systems: the atmosphere biosphere, lithosphere, and hydrosphere

b. The interaction of Earth's physical systems

c. The spatial variation in the consequences of physical processes across Earth's surface

8. The characteristics and spatial distribution of ecosystems on earth's surface: *9–12 Targets*

   a. The distribution and characteristics of ecosystems

   b. The biodiversity and productivity of ecosystems

   c. The importance of ecosystems in people's understanding of environmental issues

Human Systems

9. The characteristics, distribution, and migration of human populations on earth's surface: *9–12 Targets*

   a. Trends in world population numbers and patterns

   b. The impact of human migration on physical and human systems

10. The characteristics, distribution, and complexity of earth's cultural mosaics: *9–12 Targets*

    a. The impact of culture on ways of life in different regions

    b. How cultures shape the character of a region

    c. The spatial characteristics of the processes of cultural convergence and divergence

11. The patterns and networks of economic interdependence on earth's surface: *9–12 Targets*

    a. The classification, characteristics, and spatial distribution of economic systems

    b. How places of various size function as centers of economic activity

    c. The increasing economic interdependence of the world's countries

12. The processes, patterns, and functions of human settlement: *9–12 Targets*

    a. The functions, sizes, and spatial arrangements of urban areas

    b. The differing characteristics of settlement in developing and developed countries

    c. The processes that change the internal structure of urban areas

    d. The evolving forms of present-day urban areas

13. How the forces of cooperation and conflict among people influence the division and control of earth's surface: *9–12 Targets*

    a. Why and how cooperation and conflict are involved in shaping the distribution of social, political, and economic spaces on Earth at different scales

    b. The impact of multiple spatial divisions on people's daily lives

    c. How differing points of view and self-interests play a role in conflict over territory and resources

Environment and Society

14. How human actions modify the physical environment: *9–12 Targets*

    a. The role of technology in the capacity of the physical environment to accommodate human modification

    b. The significance of the global impacts of human modification of the physical environment

    c. How to apply appropriate models and information to understand environmental issues

15. How physical systems affect human systems: *9–12 Targets*

    a. How changes in the physical environment can diminish its capacity to support human activity

    b. Strategies to respond to constraints placed on human systems by the physical environment

    c. How humans perceive and react to natural hazards

16. The changes that occur in the meaning, use, distribution, and importance of resources: *9–12 Targets*

    a. How the spatial distribution of resources affects patterns of human settlement

    b. How resource development and use change over time

    c. The geographic results of policies and programs for resource use and management

17. How to apply geography to interpret the past: *9–12 Targets*

    a. How processes of spatial change affect events and conditions

    b. How changing perceptions of places and environments affect the spatial behavior of people

    c. The fundamental role that geographical context has played in affecting events in history

18. How to apply geography to interpret the present and plan for the future: *9–12 Targets*

    a. How different points of view influence the development of policies designed to use and manage Earth's resources

    b. Contemporary issues in the context of spatial and environmental perspectives

    c. How to use geographic knowledge, skills, and perspectives to analyze problems and make decisions

Appendix F

What's on the Web

National Aeronautics and Space Administration

300 E Street SW,
Washington, DC 20024-3210
(202) 358-0000
Web address: www.nasa.gov

Web site: NASA was formed in 1958 amid the Cold War. The headquarters manages flight centers and research centers and all other NASA-related facilities around the country. This site provides a broad overview of the scope of NASA's mission. Visitors can find the latest news about NASA as well as learn about the agency's five strategic enterprises—aerospace technology, biological and physical research, earth science, human exploration and development of space, and space science. There is also information about the agency's commercial technology-transfer programs, which seek to find ways to push technologies made inside NASA facilities out to the commercial marketplace. The agency's annual performance report can be found here, as well as links to its annual budget. Also available are descriptions of the many ongoing NASA missions, from the International Space Station project to the Space Shuttle and Hubble Space Telescope. Information about past missions can also be accessed through this portal—manned missions including the Mercury and Gemini programs of the early 1960s, and unmanned missions such as the Mars Polar Lander. There are links also to NASA's extensive educational resources, as well as NASA's Spanish-language pages.

The Robert C. Byrd National Technology Transfer Center

Wheeling Jesuit University
316 Washington Ave.

Wheeling, WV 26003
(800) 678-6882
Web address: http://www.nttc.edu/

Web site: The center, created by Congress in 1989, works to connect U.S. businesses with the technologies developed by federally funded research produced by NASA and other government agencies, such as the Environmental Protection Agency, the Department of Veterans Affairs and the Missile Defense Agency. The site puts a spotlight on scores of the new technologies developed. It is a wide-ranging list that includes breakthroughs in treating illnesses such as diabetes or Alzheimer's to computer software and hardware, new composite metals and even tools, such as a new kind of torque wrench developed for the Space Shuttle's solid rocket boosters. The site also is the center's main tool for marketing its consulting services, which include helping companies find specific technologies, training their employees and giving them market-strategy and manufacturing advice.

NASA's Regional Technology Transfer Centers

Center for Technology Commercialization (CTC)
1400 Computer Drive
Westborough, MA 01881-5043
(508) 870-0042
Web address: www.ctc.org

Web site: The CTC is NASA's Northeast Regional Technology Transfer Center (RTTC), covering the six New England States plus New York and New Jersey. It is a nonprofit company with seven satellite offices, and acts as a conduit for the transfer of NASA technologies to the private sector. The site gives a good overview on what RTTCs do, and also describes the CTC's particular focus in two areas: public safety and helping small businesses. The site links to the CTC's Public Safety Technology Center, which tries to help small law-enforcement agencies around the country find and use advanced technologies. The site also describes NASA's Business Outreach Program, which offers business-development advice with a special focus on helping small businesses as well as minority- and women-owned businesses.

The Technology Commercialization Center

12050 Jefferson Avenue, Suite 340
Newport News, VA 23606
(757) 269-0025
Web address: www.teccenter.org

Web site: The center, founded in 1999, serves Mid-Atlantic states of Delaware, Maryland, Pennsylvania, Virginia, and West Virginia. Its Web site puts a special emphasis on technologies developed at the nearby Langley Research Center in Hampton, VA, which is NASA's Center of Excellence for Structure and Materials with a primary mission of research in aeronautics and space technology. The site offers downloadable files on some of the Langley-developed technologies, and gives specific examples of some of the successes businesses have had in creating new products based on Langley-developed technologies.

Southeast Regional Technology Transfer Center

216 O'Keefe Building
Atlanta, GA 30332-0640
(800) 472-6785
Web address: www.edi.gatech.edu/nasa/

Web site: This center serves the nine-state region of North Carolina, South Carolina, Kentucky, Tennessee, Georgia, Florida, Alabama, Mississippi and Louisiana. It supports three NASA field centers: the Kennedy Space Center in Florida, the Marshall Space Flight Center in Alabama and the John C. Stennis Space Center in Mississippi. Those centers have particular areas of expertise and the Web site includes about two dozen specific new technologies that originate at the centers. These sample technologies give a good sense of how far NASA goes to push its discoveries to the marketplace. Each includes a fairly technical description as well as a list of possible commercial applications. The site also lets visitors link to the commercial technology transfer Web sites of the three field centers. The site includes a downloadable "handbook" for businesses that explains how to license NASA-developed technologies.

Great Lakes Industrial Technology Center (GLITeC)

20445 Emerald Parkway Drive, S.W.
Suite 200
Cleveland, OH 44135
(216) 898-6400
Web address: www.glitec.org

Web site: The center covers a six-state region—Minnesota, Wisconsin, Illinois, Indiana, Ohio and Michigan—and is affiliated with state-level technology development groups. The site has a nice Q&A explanation of what businesses can get from regional technology transfer centers. It highlights some of the technologies that come from the Glenn Research Center at Lewis Field. (Glenn's expertise is in areas such as aeropropulsion technologies, aerospace power, microgravity science, electric propulsion, and communications technologies.) The Web site gives specific examples of how companies took technologies in these areas and developed commercial products. It also links to some regional programs that NASA supports, such as the Glenn Garrett Morgan Commercialization Initiative, for small, minority-owned and women-owned businesses, and the Lewis Incubator for Technology, which serves startup companies in Ohio.

Mid-Continent Technology Transfer Center (MCTTC)

301 Tarrow
College Station, TX 77843-8000
(409) 845-8762
Web address: www.mcttc.com

Web site: The MCTTC serves a huge, 14-state region encompassing Texas, New Mexico, Oklahoma, Arkansas, Kansas, Missouri, Iowa, Colorado, Utah, Nebraska, Wyoming, South Dakota, North Dakota and Montana. The center is part of the Texas Engineering Extension Service, which is a member of the Texas A&M University system. It reports directly to NASA's Johnson Space Center in Houston. The Web site's content reflects the area the center covers—there are links to a diverse group of technology research centers, such as a laboratory in Montana backed by the Department of Agriculture, or the Sandia National Laboratories in New Mexico, which specializes in securing the nation's nuclear weapons stockpile. The Web site also has downloadable versions of the MCTTC's quarterly publication *TechBridge*. Each issue outlines a handful of thoroughly screened technologies available for licensing.

Far West Regional Technology Transfer Center

Far West RTTC
University of Southern California
3716 S. Hope St., Suite 200
Los Angeles, CA 90007-4344
(213) 743-2353
Web address: www.usc.edu/dept/engineering/TTC/NASA
/index.html

Web site: The FRRTTC is an Engineering Research Center within the School of Engineering at the University of Southern California. It covers an eight-state region that includes Alaska, California, Arizona, Washington, Oregon, Hawaii, Idaho and Nevada. Its Web site offers information about technologies from the three NASA field centers in California—the Jet Propulsion Laboratory, the Ames Research Center and the Dryden Flight Research Center. One of the site's best offerings is a "Learning Center" that offers downloadable PowerPoint presentations (also PDFs and HTML) on a variety of tech-transfer and business management topics. The site also has an 11-page "book" with short profiles of specific companies in the West region that were able to get new technologies or market advice from the FRRTTC.

Commercial Space Centers

(There are seventeen commercial space centers around the country, often affiliated with state universities and supported by NASA. Each has a particular area of expertise. A key part of their overall mission is to help move space-related discoveries and technologies into the private sector by working directly with the private sector.)

BioServe Space Technologies
University of Colorado—Boulder
Aerospace Engineering Sciences
429 UCB
Boulder, CO 80309-0429
(303) 492-4010
Web address: http://www.colorado.edu/engineering/BioServe or
http://www.ksu.edu/bioserve/

Web site: A place to learn about some of the custom-built hardware that makes space experiments possible. BioServe is shared by the University of

Colorado and Kansas State University. Its three main research areas are agriculture, biomedicine and biotechnology. The Web site describes some of the BioServe experiments that have flown in the payloads of missions of the Space Shuttle, aboard the *Mir* space station and the International Space Station. There are also descriptions and images of hardware such as containers for growing cell cultures and plants or for housing animals and insects. The University of Colorado has a graduate-level program in bioastronautics and microgravity sciences related to BioServe's research.

Center for Advanced Microgravity Materials Processing (CAMMP)

Department of Chemical Engineering
342 Snell Engineering
Boston, MA 02115
(617) 373-7910
Web address: http://www.dac.neu.edu/cammp/

Web site: This center, established in 1997 at Northeastern University, focuses exclusively on research into new kinds of materials. In fairly technical language the Web site describes some of the kinds of materials and research done by the center, into areas such as chemical sensors, selective membranes, gas storage devices, and molecular electronics. For the very technically oriented visitor the site offers a kind of virtual tour of the center's 3,000-square-foot research laboratory, including photos and data about the lab's equipment—such as a scanning electron microscope and a state-of-the-art analytical X-ray facility with two independent diffractometers. CAMMP experiments are also carried out in the microgravity environment of the International Space Station. Finally, the Web site includes a database with some of the real-world applications that materials science has produced, such as environmental cleanup materials.

Center for Biophysical Science and Engineering (CBSE)

Center for Biophysical Science and Engineering
University of Alabama at Birmingham
MCLM 262
1530 3rd Avenue South
Birmingham, AL 35294-0005
(205) 934-5329
Web address: http://www.cbse.uab.edu/

Web site: This center specializes in the study of biological macromolecules and uses "structure-based drug-design methodology" to development new drugs for treating disease. Housed at the University of Alabama at Birmingham, the center supports its biophysical research with engineering expertise needed to design specialized equipment. The Web site gives visitors a way to explore in detail the workings of a sophisticated multidisciplinary research center, with information on the facilities, research and the specialties of engineering, biotechnology and structural biology. The site also has links to several companies that have been formed based on research at the center.

Center for Commercial Applications of Combustion in Space (CCACS)

CCACS
Colorado School of Mines
1500 Illinois St.
Golden, CO 80401-1887
(303) 384-2091
Web address: http://www.mines.edu/research/ccacs/

Web site: Like its title says, the center is about combustion, and finding new and efficient ways to use the process to turn combustible materials such as fossil fuels into energy. The Web site gives an overview of the center's four research areas: combustors, fire safety and prevention, advanced materials, and sensors and controls. The site also has a collection of links to other combustion-related research efforts at universities, NASA field centers and the International Space Station. The CCACS is based at the Colorado School of Mines in Golden, CO.

Center for Satellite and Hybrid Communication Networks

University of Maryland
A.V. Williams Building 115
College Park, MD 20742
(301) 405-7900
Web address: http://www.isr.umd.edu/CSHCN/

Web site: This center is a research center within the Institute for Systems Research and the A. James Clark School of Engineering at the University of Maryland. Its mission is to develop hybrid networks that link satellite and wireless systems with cellular, cable, Internet, and telephone networks. The center has done research for NASA as well as the Defense Advanced Research Project Agency (DARPA). The site briefly describes those research areas; the site's strongest resource is its Publications and Presentations pages (with titles such as "Next Generation Satellite Systems for Aeronautical Communications") from which readers can view or download formal presentations on specific communications projects, presented by scholars and private-sector researchers.

Center for Space Power

Texas A&M University
Wisenbaker Building, Room 223
College Station, TX 77843-3118
(979) 845-8768
Web address: http://engineer.tamu.edu/tees/csp/

Web site: The site describes the research efforts to create new ways to produce power in space, both for NASA missions and space-power related commercial ventures. As described briefly on this site, the center does research in about half-a-dozen fields such as solar power, photovoltaic conversion, and developing new kinds of batteries.

Center for Space Power and Advanced Electronics

Space Research Institute
231 Leach Center
Auburn University, AL 36849
(334) 844-5894
Web address: http://spi.auburn.edu/cspae.htm

Web site: The center, part of Auburn University's Space Research Institute, studies ways to make electronic devices and power supplies that are more powerful and can last longer in the harsh environment of space. It works closely with the aerospace industry in three areas—high-temperature silicon-carbide technologies, electronics packaging and power supplies. The Web site gives a quick description of the specialized work being done with companies such as Northrop Grumman, Boeing and United Technologies.

Commercial Space Center for Engineering (CSCE)

223 Weisenbaker Engineering Research Center
MS3118
Texas A&M University
College Station, TX 77843-3118
(979) 845-8768
Web address: http://engineer.tamu.edu/tees/csce/

Web site: The site is a how-to for industries who want to get a payload aboard the Space Shuttle or International Space Station or other orbital platforms for experimentation. The center is a division of the Texas Engineering Experiment Station, a state agency located on the campus of Texas A&M University. The Web site answers some basic questions about why space-based testing of products is necessary and outlines a three-phase process for turning an idea into a payload ready for launch. It also lists the center's experts in fields such as robotics, nanotechnology and propulsion.

Consortium for Materials Development in Space (CMDS)

Consortium for Materials Development in Space
University of Alabama in Huntsville
Research Institute Building
4701 University Drive
Huntsville, AL 35899
(256) 824-6620
Web address: http://www.uah.edu/research/cmds/index.html

Web site: The center is part of the University of Alabama in Huntsville, also the location of the Marshall Space Flight Center. The Web site gives a brief description of its mission and its effort to recruit commercial partners for research in three specific areas: Organic nonlinear optical materials for electro-optical and all-optical applications; Z-BLAN and other heavy metal fluoride glasses for optical fibers; and mercurous chloride crystals for acousto-optical devices.

Environmental Systems Commercial Space Technology Center

P.O. Box 116450
University of Florida
Gainesville, FL 32611-6450
(352) 392-7814
Web address: http://www.ees.ufl.edu/escstc/

Web site: The center focuses on developing the recycling technologies to be used in long-duration human space flight, and on finding commercial applications for those technologies. The Web site gives a clear overview of the three main focus areas: water recovery, solid waste recovery and air revitalization. The site also has a searchable library of documents, and links to businesses and other universities working on similar projects.

Food Technology Commercial Space Center

Iowa State University
2901 South Loop Drive
Suite 3700
Ames, IA 50010-8632
(515) 296-5383
Web address: http://www.ag.iastate.edu/centers/ftcsc/

Web site: An excellent site. As the name implies this center handles the research into developing food supplies and systems for missions, from 90-day missions to the International Space Station to future missions to outposts on the Moon or Mars that could last up to three years. The Web site has a downloadable PDF brochure about the center, as well as links to current and past issues of its electronic newsletter, *NASA FTCSC News.* It also shows photos and gives details about its research facilities. A unique link from the site is its "Space Food Insights" page, which gives information about the current space-food system, such as the daily recommended diet, vitamin and mineral intakes for missions from 30 days to a year.

Medical Informatics and Technology Application Center (MITAC)

P.O. Box 980480
Richmond, VA 23298

(804) 827-1020
Web address: http://www.meditac.com

Web site: A center devoted to creating ways to make the newest medical technologies available on board manned missions to space through telemetry, telemedicine and materials science. This Web site explains how the merging of telecommunications and medicine helps not only the space program, but also makes it possible to deliver medical care to remote areas on Earth. Its 'Projects' page has an extensive list of telemedicine projects MITAC has been involved in, such as telemedicine links to remote areas of Ecuador that allowed physicians in the United States to monitor and evaluate patients. The site also offers downloadable reports about telemedicine projects in places like Turkey, Russia and the Dominican Republic as well as on specific telemedicine technologies.

ProVision Technologies (PVT)

Bldg. 9313 Room 130
Stennis Space Center, MS 39529
(228) 689-8176
Web address: http://www.pvtech.org

Web site: ProVision is a nonprofit part of the Institute for Technology Development, located at the Stennis Space Center in Mississippi. Its specialty is developing biomedical uses for hyperspectral imagery. The Web site defines hyperspectral imaging and describes its applications in four areas—food safety, skin health, retinal imaging and forensics. Examples include using spectral imaging to detect salmonella in food, or to monitor skin and eye health. The site also describes the software and imaging hardware used in these research areas, and links to related research programs at hospitals and in the private sector.

Solidification Design Center (SDC)

Professor of Mechanical Engineering
Auburn University
201 Ross Hall
Auburn, AL 36849
(334) 844-5940
Web address: http://metalcasting.auburn.edu/

Web site: This center, formed in 1997 from the Center for Space Power and Advanced Electronics, also at Auburn University, studies the properties of materials used in metal casting, using the microgravity environment of the International Space Station. The Web site explains the benefits of those low-gravity experiments and the equipment developed to make them possible. There are also downloadable files of technical data on the properties of alloys studied at the SDC as well as links to research partners from academia, business and government.

Space Communications Technology Center (SCTC)

Florida Atlantic University
Boca Raton, FL 33431
(561) 297-2343
Web address: http://www.fau.edu/divdept/comtech/ctchome.html

Web site: The SCTC is part of the Communications Technology Center of Florida Atlantic University. Its mission is to develop commercial uses of digital transmission techniques developed for sending video, audio and data to the Earth via satellite. The Web site describes specific equipment developed at the center, such as a high-definition color camera or a high-resolution scanner with telemedicine capabilities. It also briefly describes the ongoing research into telecommunications technology being done with private-sector companies such as BellSouth. A handful of technically oriented research papers are available as downloadable PDFs.

Space Vacuum Epitaxy Center (SVEC)

Space Vacuum Epitaxy Center
University of Houston
Science and Research Building One
4800 Calhoun Road
Houston, TX 77204-5507
(713) 743-3625
Web address: http://www.svec.uh.edu/

Web site: One of the most technically oriented commercial space center sites. The SVEC, part of the University of Houston, creates advanced thin film materials and devices for a wide range of commercial applications, most of which are quite obscure to the average consumer. The site describes in some detail the center's work specific areas, such as developing materials for use in electronics or new infrared sensors. The site also describes the university's Wake Shield Facility, a free-flying satellite orbiting the Earth. The satellite's design creates an ultrapure vacuum in which thin films can be grown at a higher quality than is possible on Earth.

Wisconsin Center for Space Automation and Robotics (WCSAR)

Wisconsin Center for Space Automation and Robotics
College of Engineering, UW-Madison
1500 Johnson Drive
Madison, WI 53706
(608) 262-5526
Web address: http://wcsar.engr.wisc.edu/

Web site: The center is where the hardware and software that enable biotech experiments in space is designed and made. The Web site describes in plain language (with photographs) some of those technologies, such as enclosed, environmentally controlled plant-growth chambers used in missions on board the Space Shuttle and the International Space Station. The center is a leader in space-based commercial biotech research. The Web site also describes some of the specific commercial products and companies that have benefited from the center's research. In addition, the center has expertise in robotics and remote sensing. The Web site describes a few of its commercial technologies in those fields, such as an automated coffee bean picker, or a remote-controlled aircraft with infrared sensors that can fly over farm fields and measure the condition of crops and soil.

NASA COMMERCIALIZATION/TECHNOLOGY TRANSFER SITES

NASA Space Product Development

Mail Code SD12
NASA Marshall Space Flight Center
Huntsville, AL 35812
Web address: http://spd.nasa.gov/

Web site: This is NASA's Web-based element of its Space Product Development Program, which is a campaign to persuade U.S. businesses to take advantage of NASA-produced technologies. Here businesses can find plain-language descriptions of how NASA's network of field centers and commercial space centers work to smooth the way for the transfer of space-related technologies. The site explains how research in three areas—agribusiness, biotechnology and materials—is leading to new discoveries, from anti-cancer compounds grown in space to new kinds of insulation and new ways to manufacture automotive parts. Each example includes links to the commercial space centers that were involved in the research and development. The site also offers downloads of the SPD annual report, which describes in more detail the kinds of research and development—including "case studies"—being done at the commercial space centers.

NASA Technology

Web address: http://nasatechnology.nasa.gov/

Web site: This is NASA's technology portal on the Internet, a gateway to its Technology Development program. The Web site links (via tabs on the left margin) to Web pages on commercialization, technology, the development process, NASA's education outreach, and its five Strategic Enterprises that make up NASA's Strategic Plan. The main page also has links to education sites, such as the "Just For Kids" page, which has interactive activities. The "Headlines" link has a constantly rotating selection of stories about specific NASA research projects, written for the general public. There is also a link to the NASA-TV site, which offers real-time videos of NASA missions for the public, teachers and the news media. The "Calendar" link has details about upcoming NASA events. All the links offer archives.

International Space Station Commercial Development

Director, Research Integration & Product Development Division
Office of Biological and Physical Research
Code UM
NASA Headquarters
300 E St. S.W.
Washington, DC 20546
Web address: http://commercial.hq.nasa.gov/

Web site: This is NASA's entry point for businesses to find out about research and development opportunities in the microgravity and ultra-vacuum environment aboard the International Space Station (ISS). The Web site's Commercial Opportunity page includes a "user's guide" to the ISS and details about what kind of experiments the spacecraft can support. The U.S. has set aside 30 percent of its share of the ISS's research capabilities for economic development projects. There are also downloadable reports of the commercial programs of other ISS-supporting nations, such as Japan, Canada and Russia, as well as eleven European nations.

NASA Space Product Development

Mail Code SD12
NASA Marshall Space Flight Center
Huntsville, AL 35812
Web address: http://spd.nasa.gov/

Web site: The Space Product Development Office supports NASA's initiatives in encouraging businesses to take part in the commercial potential of space exploration. The Web site makes its pitch directly to businesses, touting the benefits of space research and emphasizing the NASA resources available, such as the 17 Commercial Space Centers around the U.S. In Q&A format the site explains what the benefits to business are of space exploration. The site also has a downloadable brochure in PDF format titled "Bringing the Benefits of Space Down to Earth," as well as a handful of PDF "fact sheets" on commercial products that are based on space research. For three specific areas—materials, biotechnology and agribusiness—the site has links to commercial products and NASA commercial centers.

NASA Technology

National Technology Transfer Center
316 Washington Ave
Wheeling, WV 26003
Web address: http://www.nasatechnology.com/

Web site: This is where businesses can find out more details about specific NASA technologies in a dozen categories, such as aerodynamics, software, electronics and telecommunications. The site requires a registration but there is no charge to use the site. Its "Partnering Opportunities" link gives quick descriptions of the kinds of arrangements that NASA makes when working with companies, such as shared-cost contracts or collaborative agreements, as well as a variety of licensing agreements.

NASA Commercial Technology Network

Web address: http://nctn.hq.nasa.gov/

Web site: This Web site is a one-stop shop for all of the Web sites operated by NASA's network of programs, organizations and services either sponsored by or affiliated with the Commercial Technology Division. It has links to NASA's technology resources, the Small Business Innovation Research and Small Business Technology Transfer programs along with links to several NASA publications on commercial development of space. There is also a directory of all the NASA technology transfer-related programs as well as NASA-affiliates business incubators.

NASA TechFinder

Web address: http://technology.nasa.gov/

Web site: This is NASA's searchable Web site of technologies it is ready to license to the private sector. The site also has scores of "Technology Opportunity Sheets" or TOPS, that give details on specific technologies developed at NASA field centers. The site also has a search function for software technologies, and links to stories about technologies that have successfully been transferred to the private sector.

PUBLICATIONS

Aerospace Technology Innovation

Web address: http://www.nctn.hq.nasa.gov/innovation/index.html

Web site: The online version of the bimonthly magazine published by the Commercial Technology Division at NASA. The Web site has not only the contents of the current issue but archives dating back ten years. The site also has a search function that lets users hunt through past issues by keyword or concept. The back issues can often be downloaded in a single PDF file.

NASA Spinoff

Web address: http://www.sti.nasa.gov/tto/

Web site: The online version of NASA's annual publication highlighting the "success stories" of space-related technologies. Since 1976 the publication has featured between forty and fifity commercial products that have roots in space-related research. The Web site has a searchable database that lets users search for past spin-off topics; sometimes the database has the entire article and sometimes just a one-paragraph summary with the name or the company and the supporting NASA center. The site also has downloadable versions of the annual issue dating back about five years and links to Web sites that describe spinoffs with links to the Apollo missions and the Space Shuttle program.

Gridpoints

Web address: http://www.nas.nasa.gov/About/Gridpoints/gridpoints.html

Web site: This is the Web site for the NASA Advanced Supercomputing Division's (NAS) quarterly publication, called *Gridpoints*. The site has a downloadable PDF of the current issue and past issues going back several years. The articles, written in a nontechnical and readable style, describes the cutting-edge computer applications being developed by the NAS that are part of a wide range of space exploration efforts. The site's Educational Resources page has links to HTML-format feature stories from *Gridpoint* past issues.

Tech Briefs—Engineering Solutions for Design & Manufacturing

Web address: http://www.nasatech.com/

Web site: The Web site version of NASA's engineering magazine, published since the 1970s. Published monthly, it has a circulation is 207,000. This site, like the magazine, features exclusive reports of NASA-developed innovations, written by the engineers and scientists who do the work. The site also includes downloadable PDF "Technical Support Packages" for each of the technologies for current and past issues, as well as a Library link to past tech briefs. The articles usually feature diagrams and images of the technologies, as well as contact information. The site also has a searchable database with more than 6,000 technologies available for licensing, and a link for downloading free NASA software. The magazine is a joint publishing venture of NASA and Associated Business Publications of New York City.

SpaceDaily

Web address: http://www.spacedaily.com/

Web site: A kind of online newspaper about space-related events and technologies. Readers of this Web site will find everything from the latest updates on launches by national space programs around the globe, to opinion columns and events related to military space-related weapons systems. Much of its content comes from Agence France-Press, a French news service. There are also reports from the European Space Agency. The site also offers a free newsletter, SpaceDaily Express, that is sent via e-mail after registration.

The Space Comics Homepage

Web address: http://www.jsc.nasa.gov/er/seh/ApoCom1.html

Web site: A link from the Space Educators Handbook produced at the Johnson Space Center in Texas. Comic titles include subjects such as "What Really Happened to Apollo 13" and "The Charles Lindbergh Story." There is also a colorful comic called "Aero & Space," done by the graphics group at NASA's Langley Research Center.

Air & Space magazine

Web address: http://www.airspacemag.com/

Web site: The online version of the Smithsonian Air & Space Museum's magazine *Air&Space*, written for a general-interest audience. The site has the contents of the print version of the magazine, plus Web-based articles and "supplemental" information, such as downloadable 3-D images. Its "QT Sightings" link has Quicktime movies of rocket launches, the reentry of a Gemini capsule, even of a plane flying through the Eiffel Tower's arch. The site also has an archive of past issues dating back to 1986.

EDUCATION

NASA Explores

Web address: http://NASAexplores.com/

Web site: NASAexplores is a Web-based education resource that provides free weekly K–12 articles and lesson plans on ongoing NASA projects. The lessons are downloadable as PDFs or printable from the Web site. The program is managed by the Marshall Space Flight Center Education Programs department. The site also has a "Trivia" link that lets users search for trivia on a particular subject or review the weekly trivia articles that date back more than a year. Users can also subscribe to the site's free e-mail notice sent out weekly with abstracts of that week's articles and brief descriptions of the lesson plans and activities. For teachers, the "Resources" page has links to a variety of NASA's education-specific programs, such as its Educator Resource Center Network. This is an excellent link for teachers.

NASA Quest

Web address: http://quest.arc.nasa.gov/

Web site: An educational Web site affiliated with the Ames Research Center. Touted as a chance to "meet the people of NASA and look over their shoulders," the site has an extensive amount of offerings. There is a Q&A link that lets students in K–12 ask questions and get answers from experts on NASA science, math and technology content. The link also has a searchable database. There are profiles of NASA experts, lessons for K–12, audio and video programs via the Internet. There is also a link called "Women of NASA" which lets visitors to the page find out about women at NASA with careers in math, science and engineering. The site also has links to three major areas of exploration: astrobiology, aerospace and space exploration. All the links have search engines that can be tailored to a specific subject, person, media format or grade level. Visitors can also sign up for e-mailed updates on upcoming events.

NASA Spacelink

Web address: http://spacelink.nasa.gov

Web site: A good place for students and teachers to get to know NASA. Spacelink is produced by the agency's education division. It is an electronic library with current information about aeronautics and space research, and is designed for students and teachers. It has a searchable database and an alphabetic index. There are links to on-line instructional materials for all grade levels that include instructions for building model spacecraft, space trivia, and how to become an astronaut. The Multimedia link has interactive media that let visitors simulate a launch or dock with the International Space Station, or take virtual tours of space facilities. The Projects page has quick links to popular NASA sites such as the Galileo Mission to Jupiter and the Hubble Space Telescope.

NASA Education Program

Web address: http://education.nasa.gov/

Web site: The place for educators to learn about NASA's overall education initiatives. The Web site has information about educator workshops and fellowship programs, enrichment programs for rural and urban educators, and other offerings. The Electronic Field Trips link allows "virtual visits" to NASA sites and interactive videoconferencing through a program called NASA LIVE. The Web site also has links to the education offices of the NASA field centers, and has a state-by-state listing of contact information for educator resource centers. The Site Map page has an extensive listing of links to many of NASA's education-related resources, as well as a calendar of education-related events.

The Challenger Center for Space Science Education

1250 North Pitt St.
Alexandria, VA 22314
(703) 683-9740
Web address: http://www.challenger.org/

Web site: The Challenger Center is a not-for-profit organization founded by family members of the seven crew members who died the January 1986 Space Shuttle accident. The center has a network of 26 learning centers around the United States, as well as in Canada and the United Kingdom. The learning centers are two-room simulators where students can work on simulated missions such as a voyage to Mars or to the Moon. The Web site has a video tour of a learning center. For teachers, the site has downloadable lesson plans and supporting material on a variety of space-related subjects—comets, the International Space Station, the solar system are a few.

Students for the Exploration and Development of Space

SEDS-USA
MIT Room W20-445
77 Massachusetts Ave.
Cambridge, MA 02139-4307
(888) 321-7337
Web address: http://www.seds.org/

Web site: SEDS is an independent, student-based network founded in 1980 at MIT and Princeton University to promote the exploration and development of space. The Web site has links to the chapters at universities around the United States and details about starting a chapter. There are also pages on a variety of space-related subjects, including technology transfer, images from the Hubble Space Telescope and a detailed description on the search for extraterrestrial life (including an audio example of what sort of radio signal scientists are listening for). The site has a newsgroup for the SEDS national office. Its Resources page has links to hundreds of space-related sites, many targeted at students at a variety of grade levels.

National Air and Space Museum

7th and Independence Ave., SW
Washington, DC 20560
(202) 357-2700
Web address: http://www.nasm.si.edu/

Web site: Visitors to the Web site can get a taste of the highly popular museum. The site has images and details about the latest exhibits and the museum's facilities, such as the Albert Einstein Planetarium or the IMAX Theater. Ticket information is also available. Its Online Activities page offers items such as the "Explore the Universe Cyber Center," a simulated, interactive research center that lets visitors store and retrieve information for up to thirty days. The are also lesson plans in the "How Things Fly" page.

For Kids Only: Earth Science Enterprise

Web address: http://kids.earth.nasa.gov

Web site: A kid-friendly Web site that describes how NASA studies a variety of subjects—air, natural hazards, people, land and water. For each of the subjects the site links to an interactive page on topics such as continental drift or the El Nino effect that includes Flash-based graphics, quizzes and links to other NASA sites on the subject.

NASA PUBLIC OUTREACH

Science@NASA

Web address: http://science.nasa.gov/

Web site: "Direct to the people" is the slogan of this Web-based service. It is sponsored by the Science Directorate at NASA's Marshall Space Flight Center, and its mission is to spread the word about NASA's work to the public. The Web site is updated frequently and is very reader-friendly, with stories about recent space-related events such as solstices or lunar/solar eclipses, and experiments on board the International Space Station. There are also pages of the Web site focused on particular areas, such as earth science, rocketry or astronomy. The content is available in Spanish, and the site has links to non-NASA space-related sites in other languages such as Italian, Dutch, French, and Korean. Visitors can also sign up for a free subscription to Science@NASA Updates, which are e-mail notices that link back to new stories at the site.

NASA Solutions

Web address: http://nasasolutions.com/index.html

Web site: This is part of the technology transfer program at the Marshall Space Flight Center in Huntsville, Alabama, one of the most sophisticated such programs in NASA. The Web site's purpose is to spread the word about Marshall-developed technologies and facilities. There are downloadable files describing Marshall's capabilities in five areas: aerospace, earth science, manufacturing, optics and propulsion. The site's "Innovator's Corner" page explains how Marshall employees and contractors can report new technologies that might have commercial applications—as well as the kind of rewards they might reap. The "Technology Spotlight" link describes the latest Marshall technologies.

At Home With NASA

Web address: http://techtran.msfc.nasa.gov/at_home.html

Web site: A nice, reader-friendly Web site produced by the Marshall Space Flight Center's Technology Transfer Program, with stories about how space technologies are being used in everyday items. The examples come in ten categories—such as "at home," "at the mall" or "on the farm" and give short, playfully illustrated descriptions of space-related technologies that show up in everyday uses. This is an excellent place to give a general audience a simple introduction to the number of ways that space research is connected to everyday life.

MISCELLANEOUS

European Space Agency Technology Transfer Program

Web address: http://www.esa.int/technology/index.html

Web site: The European Space Agency has produced more than 150 success transfers of space technology. This Web site describes some of those successes in nine areas, such as engineering, environment, health, and software. The site also has access to the ESA's quarterly publication, *Preparing for the Future,* as well as an index to previous issues dating back to 1991.

NASA Invention of the Year

Web address: http://icb.nasa.gov/invention.html

Web site: The invention of the year is supported by the 14-member Inventions and Contributions Board, which was formed in the same Space Act of 1958 that established NASA. This Web site names the winning invention and in technical language describes the details of its development and potential applications. The site also lets visitors see all the competitors from a particular year and learn details about those technologies through downloadable zip files or videos.

Technology Transfer Society

Web address: http://www.t2s.org/

Web site: A not-for-profit organization formed in 1975, the society has nine chapters around the United States. The Web site includes downloadable versions of its bimonthly newsletter, *T' Squared* and the *Journal of Technology Transfer*, published three times a year.

The Space Technology Hall of Fame

Web address: http://www.spacetechhalloffame.org/

Web site: Since 1988, the privately run Space Foundation has recognized technologies that were originally developed through space research. Every year it picks a handful of technologies for admission to the Hall of Fame; the Web site describes those technologies that have been selected as well as the criteria used in their selection. The site also has a "Space Challenge" questionnaire about space-related technologies and a link to the foundation's "certified products" Web site.

Appendix G

Glossary

Source for many of these and other Earth Science terms: *Looking at Earth from Space, Glossary of Terms.* Greenbelt, Md.: NASA Office of Mission to Planet Earth, n.d. http://eartheducator.gsfc.nasa.gov/Glossaries/. Accessed 7 June 2003.

air pollution. A measure of air quality. Contaminants in the air come mainly from manufacturing industries, electric power plants, exhaust from automobiles, buses, and trucks.

altimeter. An instrument used to measure the altitude of an object above a fixed level.

Ames Research Center (ARC). Located at Moffett Field, California, ARC is active in aeronautical research, life sciences, space science, and technology research. The Center houses the world's largest wind tunnel and the world's most powerful supercomputer system.

analog. A continuously variable signal as opposed to a digital or discretely variable signal.

anthrax. An airborne pathogen that has become synonymous with bioterrorism. Anthrax infection can occur naturally in hooved animals and can be spread to people following contact with infected animals or contaminated animal products, or can occur as a result of intentional release of anthrax spores as a biological weapon. A NASA spin-off technology is one of the tools being used to clear air of anthrax spores.

anti-icing. Ant-icing prevents ice from building up. NASA's efforts can help improve aircraft safety, railroad safety, and road safety.

Apollo 13. During this April 1970 Apollo mission, an oxygen tank explosion in the service module forced the three-man crew of James Lovell, Jr., Fred

W. Haise, Jr., and John L. Swigert, Jr., to use the lunar module rather than their command module for the return trip. The dramatic events were the subject of a book by Lovell, later made into the movie *Apollo 13.*

Apollo-Soyuz Project. The first international space mission in July 1972. *Apollo 18* astronauts and the crew of a Russian *Soyuz* mission linked spacecraft for 44 hours. This was the final mission in Project Apollo.

array. A systematic arrangement or grouping.

artificial intelligence. The branch of computer science that attempts to program computers to respond as if they were thinking—capable of reasoning, adapting to new situations, and learning new skills. Examples of artificial intelligence programs include those that can locate minerals underground and understand human speech.

atmosphere. The air surrounding the Earth. The atmosphere, composed mainly of nitrogen and oxygen with traces of carbon dioxide, water vapor, and other gases, acts as a buffer between Earth and the sun.

bidirectional telemetry. Back and forth transmission of data. NASA's research in telemetry has advanced programmable, implantable medical devices.

biofeedback. A technique used to help people relax and manage stress. Sensors connecting a computer user to the computer can help people to train themselves to recreate certain responses. Biofeedback techniques developed for astronauts are being used with pilots and with children and adults to improve attention span.

biome. A well-defined terrestrial environment, such as a tropical forest, tundra or desert.

bionic. A system or instrument modeled after a living organism; an artificial replacement of a body part.

BIO-Plex (Bioregenerative Planetary Life Support Systems Test Complex). An experimental lab for advanced planetary life support at Johnson Space Center. It will be a state-of-the-art eight-chamber test facility. Human habitation in the BIO-Plex is scheduled to begin in 2003, with the first test lasting 120 days and the second test to last twice as long.

biopsy. A surgical procedure to further diagnose a tissue sample.

biosphere. Part of the Earth system in which life can exist.

canopy. The layer formed naturally by the leaves and branches of trees and plants.

carbon dioxide (CO_2). A minor but very important component of the *atmosphere*, carbon dioxide traps *infrared radiation*. Atmospheric CO_2 has increased about 25 percent since the early 1800s, with an estimated increase of 10 percent since 1958 (burning fossil fuels is the leading cause of increased CO_2, deforestation the second major cause).

cardiac pacemaker. An implantable device that delivers a steady pulse to help speed up an abnormally slow heartbeat.

Chandra X-ray Observatory. Chandra is NASA's most sophisticated X-ray observatory ever. It was launched on July 23, 1999. Chandra can observe X-rays from high-energy regions of the universe, such as the remnants of exploded stars. High-resolution images from Chandra often reveal important new features for space scientists to study.

Charge Coupled Devices (CCD). Silicon chips that convert light into electronic images.

condensation. Change of a substance to a denser form, such as gas to a liquid. The opposite of evaporation.

defibrillator. A device used following ventrical fibrillation when the heart stops pumping blood; death or brain damage can occur within minutes. Defibrillators can be external, such as those used in hospitals and by emergency personnel; NASA technology contributed to an implantable defibrillator, which can deliver a shock to the patient's heart when an abnormally fast heartbeat is detected.

deicing. Deicing is the process of removing ice. NASA has an entire branch researching deicing methods, such as heat, or agents, such as liquids. NASA's efforts can help improve aircraft safety, railroad safety, and road safety.

diamond-like carbon. A thin coating material applied to plastic or glass lenses to make them more scratch-resistant.

Doppler effect. The apparent change in frequency of sound or light waves, varying with the relative velocity of the source and the observer. If the source and observer draw closer together, the frequency is increased. Named for Christian Doppler, Austrian mathematician and physicist (1803–1853).

Doppler radar. The weather radar system that uses the Doppler shift of radio waves to detect air motion that can result in tornadoes and precipitation, as previously developed weather radar systems do. It can also measure the speed and direction of rain and ice, as well as detect the formation of tornadoes sooner than older radars.

drag. Resistance to motion through the air.

Dryden Flight Research Center. The Dryden Flight Research Center at Edwards Air Force Base, California, formerly part of ARC, became a separate entity March 1994. Since the 1940s, this Mojave Desert site has been a testing ground for high-performance aircraft and is one of two prime landing sites for the Space Shuttle.

ecology. The study of interrelationships between living organisms and their environments.

ecosystem. A natural unit including living and nonliving parts that interact and produce a stable system through exchange of materials.

Electro-Expulsive Separation System (EESS). The EESS is known as the "ice zapper," a system developed at Ames Research Center. An electrical current causes minute movement in parallel copper wires wrapped and bonded to aircraft wings; the movement is enough to break up ice as it forms.

electromagnetic spectrum. The entire range of radiant energies or wave frequencies from the longest to the shortest wavelengths. The spectrum usually is divided into seven sections: radio, microwave, infrared, visible, ultraviolet, X-ray, and gamma-ray radiation.

emission. Substances discharged into the air.

ergonomics. The study of problems encountered by people in their environments; the science that works to help people adapt to the physical demands of their work environment.

ethylene. Ethylene is a natural hormone that is produced by plants as they ripen. Too much of it causes plants to whither and spoil. A system to convert ethylene into carbon dioxide and water will be used on the International Space Station greenhouse and is used to keep fruits and vegetables fresher longer here on Earth.

evaporation. The change from liquid to vapor or gas form. Evaporation is the opposite of condensation.

fairing. An auxiliary structure on the external surface of a vehicle, such as an aircraft, that serves to reduce drag.

far infrared. Electromagnetic radiation, longer than the thermal infrared, with wavelengths between about 25 and 1000 micrometers.

filter. A filter selectively passes desired frequencies and removes undesired ones. Filters can be used for sound or light.

fog. Fog forms when water vapor is condensed. NASA discovered that fog could be a problem for astronauts in space as sweat caused formation of fog on helmet faceshields thick enough to obscure vision. Anti-fog sprays were developed as a result.

Food and Drug Administration (FDA). The Food and Drug Administration is part of the Public Heatlh Service of the U.S. Department of Health and Human Services. It is the regulatory agency responsible for ensuring the safety of medical devices and drugs and approving any new technology before it can go on the market in the United States.

force-feedback. A technology that puts the sense of feel into training simulators for pilots and surgeons, as well as computer games. A touch-enable mouse even brings force-feedback technology to desktop computers.

freeze-dried foods. Foods that have been dehydrated; rehydrating or adding water reconstitutes the food. Freeze-dried foods are much lighter and smaller.

geostationary. Geostationary describes the orbit in which a satellite is always in the same position over the rotating Earth. The satellite travels around the Earth in the same direction, appearing to stay in one place.

G-Force. The measure of gravitational pull. On Earth, G-force is 1.0; during space flight, it drops to 0.001.

glider. A glider is an aircraft without an engine.

Global Positioning Satellite (GPS). A constellation of 24 satellites created by the U.S. Department of Defense to provide the armed forces with all-weather navigation capabilities. A GPS receives combines data from three satellite points, thereby providing a precise position measurement. NASA uses GPS in its Space Shuttle and Earth-monitoring flights as well as on the International Space Station.

Goddard Space Flight Center (GSFC). Goddard was NASA's first major scientific laboratory devoted entirely to the exploration of space. Located in Greenbelt, Maryland, GSFC's responsibilities include design and construction of new scientific and applications satellites, as well as tracking and communication with existing satellites in orbit. GSFC is the lead center for the Earth Observing System, a key element of Mission to Planet Earth. GSFC also directs operations at the Wallops Flight Facility on Wallops Island, Virginia, which each year launches some fifty scientific missions to suborbital altitudes on small sounding rockets.

greenhouse gases. Greenhouse gases include carbon dioxide, methane, nitrous oxide, chlorofluorocarbons, and water vapor. Carbon dioxide, methane, and nitrous oxide have natural sources; chlorofluorocarbons are produced by industry. As these gases affect the chemistry of Earth's atmosphere, it could increase Earth's average temperatures and a number of plant and animal species could be threatened with extinction.

hang glider. The hang glider or parawing was at one point considered as a device to bring space payloads back to earth. Research and development by NASA on the concept helped push hang gliding into popular sport.

haptics. The science of feel. NASA uses haptics in its simulated training program so astronauts can feel the resistance they might encounter during procedures in space.

heart rate. Scientists track a variety of vital signs of astronauts during flight. During an early Mercury mission, they found that astronaut Gordon Cooper's heart rate remained around 55 beats per minute while he was sleeping and was racing at 184 beats per minute during reentry. NASA continues to work on ways to gather more information about the effects of space flight on the body with less invasive measuring devices.

Hubble Space Telescope. Launched in 1990, the Hubble Space Telescope provides ten times the resolution of any earth-bound telescope, providing scientists with precise images and data.

hydroponics. Hydroponics is the science of growing plants in a nutrient-rich solution or moist material other than soil.

hypersonic. Velocities of Mach number five or above.

imaging. Creating a pictorial representation of data, especially data acquired by satellite systems. An image is not a photograph, but is composed of two-dimensional grids of individual picture elements or pixels.

infrared. Infrared is the portion of the electromagnetic spectrum where wavelength spans the region from about 0.7 to 1000 micrometers (longer than visible radiation, shorter than microwave radiation). In the visible and near-infrared regions, vegetation cover and biological properties of surface matter can be measured. In the mid-infrared region, geological formations can be detected. In the far infrared, emissions from Earth's atmosphere and surface can be measured. Infrared detectors do not rely on visible light, so they can be used day or night, in any conditions.

insulin. Insulin is a hormone required to convert sugars and starches from the food we eat into the energy we need. NASA technology has led to the development of two kinds of insulin pumps—external and implantable—that allow insulin-dependent diabetics to infuse insulin into their bodies without daily injections.

International Space Station (ISS). The ISS is the largest and most complex scientific project in history. It is more than four times as large as the Russian Mir space station. The project is supported by sixteen nations, including eleven European nations, the United States, Canada, Japan, Russia and Brazil.

Jet Propulsion Laboratory (JPL). Located in Pasadena, California, JPL is operated under contract to NASA by the California Institute of Technology. Its primary focus is the scientific study of the solar system, including exploration of the planets with automated probes. Most of the lunar and planetary spacecraft of the 1960s and 1970s were developed at JPL. JPL also is the control center for the worldwide Deep Space Network, which tracks all planetary spacecraft.

Lyndon B. Johnson Space Center (JSC). Johnson Space Center, located between Houston and Galveston, Texas, is the lead center for NASA's human space flight program. JSC has been Mission Control for all piloted space flights since 1965, and now manages the Space Shuttle program. JSC's responsibilities include selecting and training astronauts, designing and testing vehicles and other systems for piloted space flight, and planning and executing space flight missions. The center has a major role in developing the Space Station. In addition, JSC directs operations at the White Sands Test Facility in New Mexico, which conducts Shuttle-related tests. The nearby White Sands Missile Range also serves as a back-up landing site for the Space Shuttle.

Kennedy Space Center (KSC). Located near Cape Canaveral, Florida, KSC is NASA's primary launch site. The Center handles the preparation, integration, checkout, and launch of space vehicles and their payloads. All piloted space missions since the Mercury program have been launched from here, including Gemini, Apollo, Skylab, and Space Shuttle flights. KSC is the Shuttle's homeport, where orbiters are serviced and outfitted between missions, and then assembled into a complete Shuttle "stack" before launch. The Center also manages the testing and launch of unpiloted space vehicles from an array of launch complexes, and conducts research programs in areas of life sciences related to human space flight.

Landsat. A land remote-sensing satellite. The United States' system is called Landsat. French system is called SPOT.

Langley Research Center (LaRC). Oldest of NASA's field centers, LaRC is located in Hampton, Virginia, and focuses primarily on aeronautical research. Established in 1917 by the National Advisory Committee for Aeronautics, the Center currently devotes two-thirds of its programs to aeronautics, and the rest to space. LaRC researchers use more than forty wind tunnels to study improved aircraft and spacecraft safety, performance, and efficiency.

laser (light amplification by stimulated emission of radiation). An instrument that produces predictable pulses of light.

Lewis Research Center (LeRC). Lewis Research Center, located outside Cleveland, Ohio, conducts a varied program of research in aeronautics and space technology. Aeronautical research includes work on advanced materials and structures for aircraft. Space-related research focuses primarily on power and propulsion. Another significant area of research is in energy and power sources for spacecraft, including the Space Station, for which LeRC is developing the largest space power system ever designed.

light emitting diodes (LED). The light generated by LED has a near-infrared longer wavelength. LEDs are being used to grow plants in space; they are also being used as a substitute for laser light in some cancer surgeries and in other medical therapies.

lunar landing. The first lunar landing occurred on July 20, 1969.

lunar module. A specially designed space vehicle designed to fly in a vacuum, which would allow two of the three crewmembers to float to the Moon's surface for further exploration, while the third stayed inside the lunar orbiter.

Mars Pathfinder. *Mars Pathfinder* landed on Mars on July 4, 1997, demonstrating for the first time the ability of engineers to deliver a semi-autonomous roving vehicle capable of conducting science experiments on the surface of another planet. *Mars Pathfinder* transmitted more than 16,000 images from the lander and 550 images from the rover and more than 15 chemical analyses of rocks, data on winds, and weather factors.

Mars Surveyor. The *Mars Global Surveyor* was the first to be launched in a decade-long exploration of Mars by NASA. The launches, scheduled through 2005, involve orbiters, landers, rovers and probes to Mars.

George C. Marshall Space Flight Center (MSFC). The MSFC, located in Huntsville, Alabama, is responsible for developing spacecraft hardware and systems, and is perhaps best known for its role in building the Saturn rockets that sent astronauts to the Moon during the Apollo program. It is NASA's primary center for space propulsion systems and plays a key role in the development of payloads to be flown on the shuttle (such as Spacelab). MSFC also manages two other NASA sites: the Michoud Assembly Facility in New Orleans where the Shuttle's external tanks are manufactured, and the Slidell Computer Complex in Slidell, Louisiana, which provides computer support to Michoud and to NASA's John C. Stennis Space Center.

memory metal. Metal that is able to return to its original shape if deformed.

microbursts. Microbursts are a dangerous weather phenomenon. They are columns of air, cooled by thunderstorms that plummet downward at speeds up to 150 miles per hour. Pilots coming through a microburst may lose control over their aircraft because they do not recognize the rapidly changing wind conditions.

microwave. Electromagnetic radiation with wavelengths between about 1000 micrometers and one meter.

middle infrared. Electromagnetic radiation between the near infrared and the thermal infrared, about 2 to 5 micrometers.

Moderate Resolution Imaging Spectroradiometer (MODIS). MODIS is a key instrument aboard the Terra and Aqua satellites. Terra MODIS and Aqua MODIS are viewing the entire Earth's surface every one to two days, acquiring data in 36 spectral bands, or groups of wavelengths. The data improve understanding of global dynamics and processes occurring on the land, in the oceans, and in the lower atmosphere.

National Aeronautics and Space Administration (NASA). U.S. Civilian Space Agency created by Congress in 1958, NASA belongs to the executive branch of the Federal Government. NASA's mission to plan, direct, and conduct aeronautical and space activities is implemented by NASA Headquarters in Washington, D.C., and by ten major centers spread throughout the United States. The agency administers and maintains these facilities; builds and operates launch pads; trains astronauts; designs aircraft and spacecraft; sends satellites into Earth orbit and beyond; and processes, analyzes, and distributes the resulting data and information. NASA's programs of basic and applied research extend from microscopic subatomic particles to galactic astronomy. In addition to enhancing scientific knowledge, thousands of the technologies developed for aerospace have resulted in commercial applications.

National Oceanic and Atmospheric Administration (NOAA). NOAA was established in 1970 as part of the U.S. Department of Commerce, directed to ensure the safety of the general public from atmospheric phenomena and to provide the public with an understanding of the Earth's environment and resources. NOAA's two main components are the National Weather Service (NWS) and the National Environmental Satellite, Data, and Information Service (NESDIS).

near infrared. Electromagnetic radiation with wavelengths from just longer than the visible (about 0.7 micrometers) to about two micrometers.

osteoporosis. Osteoporosis is a reduction in bone density accompanied by brittleness. The condition affects primarily—but not exclusively—postmenopausal women. Astronauts suffer a similar disuse osteoporosis in microgravity. There, bones atrophy at a rate of about 1 percent a month and could reach 40 to 60 percent.

ozone. A molecule consisting of three oxygen atoms. Ozone absorbs ultraviolet light, which makes it useful in the stratosphere because it protects life on earth from the damaging effects of this radiation. It is harmful on Earth's surface because it can damage the lung tissue of those who breathe it.

ozone hole. An area of stratospheric ozone depletion over the Antarctic continent. It typically occurs in late summer and lasts until mid-November. Scientists say this phenomenon is the result of chemical mechanisms initiated by man-made chlorofluorocarbons.

ozone layer. The level of the atmosphere that begins about 15 kilometers above earth and thins out almost completely at 50 kilometers.

ozone-measuring satellite instruments. Satellite-based ozone-measuring instruments can measure ozone by looking at the amount of ultraviolet absorption reflected from the Earth's surface and clouds. Some instruments provide data within the different levels of the atmosphere. The Total Ozone Mapping Spectrometer (TOMS) maps the total amount of ozone between ground and the top of the atmosphere.

pacemaker. An implantable device that delivers a steady pulse to help speed up an abnormally slow heartbeat.

payload. The part of the cargo not essential to flight operations. Payload can include material for scientific experiments, equipment for making repairs in space or equipment to be left behind in orbit.

pneumatic. Filled with or operated by compressed air.

Project Apollo. Project Apollo was NASA's third human space flight project. The goal was to get astronauts to the moon, and it succeeded in that goal. In total, six expeditions landed on the moon and *Apollo 13*'s planned lunar landing was aborted. Project Apollo ended in 1972.

Project Gemini. Project Gemini was NASA's human space flight project that followed Project Mercury. Gemini missions were in two-man spacecrafts. During its ten missions, the Gemini project added nearly 1,000 hours in space-flight experience. Project Gemini ended in November 1966.

Project Mercury. The first NASA human space flight project. There were six Mercury flights, totaling just over two hours in space. The first Mercury flight was on May 5, 1961; the final one was on May 15–16, 1963.

prototype. A prototype is a model or pattern. Prototypes can be full scale or miniaturized versions of new technologies.

Quantum Well Infrared Photodetectors (QWIPs). An ultrasensitive digital infrared sensor originally developed in the early 1990s to search for distant galaxies and as part of the Star Wars defense program to spot missile launches. Since then, the technology has resulted in one of the most sensitive handheld infrared cameras.

Rotational Hand Controller (RHC). A joystick-like device developed for NASA to provide more realistic training for astronauts.

sensor. A device that produces an output, usually electrical, in response to a stimulus. Sensors aboard satellites, for example, obtain information about features and objects on Earth by detecting radiation reflected or emitted in different bands of the electromagnetic spectrum.

shock. A condition resulting from ineffective circulation of blood. Shock was a concern for NASA because without gravity, blood is no longer pulled down toward the feet, and blood volume in the upper body begins to increase. As astronauts come back to Earth and the full force of gravity, blood volume in the upper body drops. NASA's development of anti-gravity suits has resulted in anti-shock garments, essential equipment in many emergency situations.

Skylab. *Skylab* was NASA's space station, designed to let astronauts stay in Earth orbit for extended periods. Three crews of three astronauts spent time in *Skylab* between 1973 and 1974. The empty *Skylab* station reentered Earth's atmosphere and burned up on July 11, 1979.

smoke detector. A NASA technology designed to help astronauts detect a fire on board the large *Skylab* space station. Because smoke does not rise in space, smoke detectors are placed in and around the ventilation system.

space pen. Developed by Fisher Space Pens, these pens contain semisolid ink that won't leak and can write in any position—in microgravity, upside down, or under water.

Space Shuttle. NASA's most recent human space flight project. Unlike earlier space capsules, the Shuttle is an airplane-like, reusable spacecraft. Space Shuttle flew its first mission in 1981.

spin-off. A technology created for one purpose and used for additional purposes. NASA has an active technology transfer initiative to move technologies designed for or developed by NASA into the private sector.

John C. Stennis Space Center (SSC). This Center, located on Mississippi's Gulf Coast, is NASA's prime test facility for large liquid propellant rocket engines and propulsion systems. The main mission of the Center is to support testing on a regular basis of the Space Shuttle's main propulsion system. SSC is responsible for a variety of research programs in the environmental sciences and for the remote sensing of Earth resources, weather, and oceans, and is the lead NASA Center for the commercialization of space remote sensing.

supersonic. Faster than the speed of sound.

telemetry. Telemetry is the transmission of data collected at a remote location over communications channels to a central station. NASA's research in telemetry has advanced programmable, implantable medical devices.

thermal infrared. Electromagnetic radiation with wavelengths between about 3 and 25 micrometers.

thrust. The forward-directed force developed in a jet or rocket engine as a reaction to the high-velocity rearward ejection of exhaust gases.

ultraviolet. The energy range just beyond the violet end of the visible spectrum. Ultraviolet radiation constitutes only about 5 percent of the total energy emitted from the sun, but it is the major energy source for the stratosphere and mesosphere, playing a dominant role in both energy balance and chemical composition. Most ultraviolet radiation is blocked by Earth's atmosphere, but some solar ultraviolet penetrates and aids in plant photosynthesis and helps produce vitamin D in humans. Too much ultraviolet radiation can burn the skin, cause skin cancer and cataracts, and damage vegetation.

virtual reality. Virtual reality allows computer users to immerse themselves into a realistic-looking computer-generated environment. The 3-D graphic environments can be explored and manipulated by the user.

visible light. That part of the electromagnetic spectrum to which the human eye is sensitive, between about 0.4 and 0.7 micrometers.

zeolite. Zeolite is crushed rock mined from ancient volcanic ash deposits. It's lightweight and more porous than ordinary soil, so it holds moisture. It is used by NASA and on Earth as a growing medium for plants. One of its characteristics is a high cation exchange capacity, meaning that plant nutrients such as nitrogen and potassium can be held to the zeolite. These nutrients are released slowly, fertilizing the roots of the plants.

zero-gravity posture. The position astronauts' bodies assume when they relax in microgravity environment. Rather than sitting up straight or laying flat, the body forms about a 128 degree angle from trunk to thigh.

Index

About the Authors

MARJOLIJN BIJLEFELD is a freelance writer and editor. She is the author of *The Gun Control Debate: A Documentary History* (Greenwood, 1997), *People For and Against Gun Control* (Greenwood, 1999), and *Food and You: A Guide to Healthy Habits for Teens* (Greenwood, 2001).

ROBERT BURKE is a business and technology writer.